AF801362

CPSS Power Electronics Series

Series Editors

Wei Chen, Fuzhou University, Fuzhou, Fujian, China

Yongzheng Chen, Liaoning University of Technology, Jinzhou, Liaoning, China

Xiangning He, Zhejiang University, Hangzhou, Zhejiang, China

Yongdong Li, Tsinghua University, Beijing, China

Jingjun Liu, Xi'an Jiaotong University, Xi'an, Shaanxi, China

An Luo, Hunan University, Changsha, Hunan, China

Xikui Ma, Xi'an Jiaotong University, Xi'an, Shaanxi, China

Xinbo Ruan, Nanjing University of Aeronautics and Astronautics, Nanjing Shi, Jiangsu, China

Kuang Shen, Zhejiang University, Hangzhou, Zhejiang, China

Dianguo Xu, Harbin Institute of Technology, Haerbin Shi, Heilongjiang, China

Jianping Xu, Xinan Jiaotong University, Chengdu, Sichuan, China

Mark Dehong Xu, Zhejiang University, Hangzhou, Zhejiang, China

Xiaoming Zha, Wuhan University, Wuhan, Hubei, China

Bo Zhang, South China University of Technology, Guangzhou Shi, Guangdong, China

Lei Zhang, China Power Supply Society, Tianjin, China

Xin Zhang, Hefei University of Technology, Heifei Shi, Anhui, China

Zhengming Zhao, Tsinghua University, Haidian Qu, Beijing, China

Qionglin Zheng, Beijing Jiaotong University, Haidian, Beijing, China

Luowei Zhou, Chongqing University, Chongqing, Sichuan, China

This series comprises advanced textbooks, research monographs, professional books, and reference works covering different aspects of power electronics, such as Variable Frequency Power Supply, DC Power Supply, Magnetic Technology, New Energy Power Conversion, Electromagnetic Compatibility as well as Wireless Power Transfer Technology and Equipment. The series features leading Chinese scholars and researchers and publishes authored books as well as edited compilations. It aims to provide critical reviews of important subjects in the field, publish new discoveries and significant progress that has been made in development of applications and the advancement of principles, theories and designs, and report cutting-edge research and relevant technologies. The CPSS Power Electronics series has an editorial board with members from the China Power Supply Society and a consulting editor from Springer.

Readership: Research scientists in universities, research institutions and the industry, graduate students, and senior undergraduates.

Xinbo Ruan · Lihong Xie · Qing Ji · Xibo Yuan

Conducted Electromagnetic Interference in Power Converters: Modeling, Prediction and Reduction

Springer

Xinbo Ruan
College of Automation Engineering
Nanjing University of Aeronautics
and Astronautics
Nanjing, China

Lihong Xie
College of Automation Engineering
Nanjing University of Aeronautics
and Astronautics
Nanjing, China

Qing Ji
College of Mechanical and Electrical
Engineering
Soochow University
Suzhou, China

Xibo Yuan
School of Electrical Engineering
China University of Mining and Technology
Xuzhou, China

ISSN 2520-8853 ISSN 2520-8861 (electronic)
CPSS Power Electronics Series
ISBN 978-981-99-9294-2 ISBN 978-981-99-9295-9 (eBook)
https://doi.org/10.1007/978-981-99-9295-9

© The Editor(s) (if applicable) and The Author(s), under exclusive license to Springer Nature Singapore Pte Ltd. 2024

This work is subject to copyright. All rights are solely and exclusively licensed by the Publisher, whether the whole or part of the material is concerned, specifically the rights of translation, reprinting, reuse of illustrations, recitation, broadcasting, reproduction on microfilms or in any other physical way, and transmission or information storage and retrieval, electronic adaptation, computer software, or by similar or dissimilar methodology now known or hereafter developed.
The use of general descriptive names, registered names, trademarks, service marks, etc. in this publication does not imply, even in the absence of a specific statement, that such names are exempt from the relevant protective laws and regulations and therefore free for general use.
The publisher, the authors, and the editors are safe to assume that the advice and information in this book are believed to be true and accurate at the date of publication. Neither the publisher nor the authors or the editors give a warranty, expressed or implied, with respect to the material contained herein or for any errors or omissions that may have been made. The publisher remains neutral with regard to jurisdictional claims in published maps and institutional affiliations.

This Springer imprint is published by the registered company Springer Nature Singapore Pte Ltd.
The registered company address is: 152 Beach Road, #21-01/04 Gateway East, Singapore 189721, Singapore

Paper in this product is recyclable.

Preface

With the development of power electronics technology, power converters have been widely used in various fields such as general industry, power system, electrified transportation, information technology, aerospace and home appliance and have also become increasingly essential in leading areas such as renewable energy generation, flexible AC/DC power transmission, electric vehicles, energy saving and environmental protection. However, the switching actions of power devices in power converters result in high changing rates in voltage (dv/dt) and current (di/dt), and the conducted emissions are generated. These emissions not only pollute the grid, but also affect the operation of nearby equipment, leading to the conducted electromagnetic interference (EMI) issues. With the applications of wide-band gap power devices and the increased power density, the electromagnetic compatibility (EMC) issue becomes more urgent to be solved.

This book aims to present the modeling, prediction and reduction of the conducted EMI in power converters, including the AC-DC rectifiers, DC-DC converters and DC-AC inverters. The detailed theoretical analysis with design examples and experimental validations is covered.

This book contains 11 chapters.

In Chap. 1, the basic concepts of EMC and EMI are briefly introduced. The generation of common-mode (CM) and differential-mode (DM) EMI in power converters is presented. Then, the research status and key issues for the conducted EMI in the power converters are reviewed, including the AC-DC rectifiers, DC-DC converters and DC-AC inverters.

In Chap. 2, the main instruments for measuring the conducted EMI are introduced, and their operation principles are explained in detail. This chapter also presents the design of EMI filters.

Chapters 3–5 focus specifically on the conducted EMI of boost power factor correction (PFC) converters.

In Chap. 3, the propagation paths for the CM and DM noises of the PFC converter is analyzed, and the DM capacitors are added at the AC and DC sides of the rectifier bridge to suppress the mixed-mode noise. After that, the CM and DM noise equivalent circuits are derived, which provide the basis for predicting the CM and DM noise spectra of boost PFC converter with different control schemes. The design of EMI filters for the boost PFC converter is also presented.

In Chap. 4, the operation modes of average-current-controlled boost PFC converter are analyzed. The drain-to-source voltage for the power switch is treated as the noise voltage source that creates the conducted EMI, and its spectrum characteristics are discussed. Then, the CM and DM noise spectra under different input voltage and load are predicted, and the worst conducted EMI spectrum is figured out. Based on the worst conducted EMI spectrum, the EMI filter design is developed.

Chapter 5 focuses on the modeling and prediction for the conducted EMI in critical conduction mode (CRM) boost PFC converter. The spectrum characteristics of the drain-to-source voltage for the power switch are analyzed, and the CM and DM noise spectra are investigated. Based on the model of EMI receiver, the peak, quasi-peak and average values of the CM and DM noises are analyzed. By deriving the maximum boundaries for the CM and DM noise, the design guidelines for the EMI filter of CRM boost PFC converters are presented.

Chapters 6–10 present the modeling and reduction of the CM noise in DC-DC converters.

In Chap. 6, a generalized lumped capacitance model of the transformer is derived, and then the CM noise model of isolated DC-DC converters. Based on the CM noise model, the equivalent noise source (ENS) concept is proposed to analyze the CM noise characteristics of basic isolated power converters. The topologies that have the capability of CM noise cancelation (CMNC) are identified, and the effects of transformer winding structure as well as the circuit configuration on the CM noise are discussed.

Chapter 7 focuses on the shielding-based approach for reducing the CM noise in isolated DC-DC converters. The shielding-winding method and the combined shielding and balancing-winding method are proposed to mitigate the displacement current flowing through the transformer inter-winding capacitance. Furthermore, the hybrid passive cancelation (HPC) method is developed to further cancel the displacement current flowing through the parasitic capacitance between the high dv/dt nodes in the primary circuit and the protective earth (PE).

In Chap. 8, the CM noise reduction for phase-shifted full-bridge (PSFB) converter is specially discussed, as this topology does not have the CMNC capability. The electric potentials of two phase-leg midpoints and the voltage across the resonant inductor are identified as the noise voltage sources that generate the CM noise, and then a combined method of symmetry circuit and passive cancelation is proposed to eliminate the impact of these noise source voltages on the CM noise effectively, and thus reduce the CM noise.

Chapter 9 explores the CM voltage cancelation (CMVC) method in DC-DC converters. This method introduces a CM voltage compensation branch connected in series with the input of power converter to cancel the CM voltage of the converter. The applications of the CMVC method in both non-isolated and isolated DC-DC converters are presented.

Chapter 10 focuses on the CM currents reduction at both the input and output sides of non-isolated DC-DC converters. The split-winding (SW) method is proposed, where the original inductor winding is split and redistributed into the input, ground and output sides for canceling the CM currents.

Chapter 11 is dedicated to the CM currents reduction at both the input and output sides of DC-AC inverters. The CMVC method is proposed, where two CM transformers are inserted at the input and output sides of the inverter for canceling the CM voltage of the inverter.

This book is essential and valuable reference for graduate students and academics majoring in power electronics. It is also beneficial for engineers involved in addressing EMC issues in aerospace power supply, server power supply, charger and drive systems in electrical vehicles, and renewable energy generation systems. Senior undergraduate students majoring in electrical engineering and automation engineering would also find this book useful.

Nanjing, China Xinbo Ruan
Nanjing, China Lihong Xie
Suzhou, China Qing Ji
Xuzhou, China Xibo Yuan

Acknowledgements

This research monograph summarizes the research work on the conducted electromagnetic interference (EMI) of power converters since 2009.

The work in this book was partially supported by the Lite-On Technology Corporation. We would like to express our sincere gratitude to this support.

GPT-3.5 was used to refine the English language in this research monograph.

It has been a great pleasure to work with the colleagues of Springer, Science Press, China, and China Power Supply Society (CPSS). The support and help from Mr. Wayne Hu (the Project Editor) and Vidyaa Shri Krishna Kumar are greatly appreciated.

July 2023

About This Book

In power converters, the switching actions of power devices generate high changing rates in voltage (dv/dt) and current (di/dt), leading to the conducted emissions. These emissions not only pollute the grid, but also affect the operation of nearby equipment, resulting in the conducted electromagnetic interference (EMI) issues. This book focuses on the modeling, prediction and reduction of the conducted EMI in power converters, including the AC-DC rectifiers, DC-DC converters and DC-AC inverters. The detailed theoretical analysis with design examples and experimental validations is covered. This monograph is not only a reference book for master students, Ph.D. students and teachers majoring in power electronics, but also a handbook for the electrical engineers working on EMC issues in aerospace power supply, server power supply, charger and drive systems in electrical vehicles, renewable energy generation systems, etc.

Contents

About the Authors

Xinbo Ruan received the B.S. and Ph.D. degrees in electrical engineering from Nanjing University of Aeronautics and Astronautics (NUAA), Nanjing, China, in 1991 and 1996, respectively.

In 1996, he joined the Faculty of Electrical Engineering Teaching and Research Division, NUAA, where he became a Professor in the College of Automation Engineering in 2002 and has been engaged in teaching and research in the field of power electronics. From August to October 2007, he was a Research Fellow in the Department of Electronic and Information Engineering, Hong Kong Polytechnic University, Hong Kong, China. From March 2008 to August 2011, he was also with the School of Electrical and Electronic Engineering, Huazhong University of Science and Technology, China. He is the author or co-author of 13 books and more than 300 technical papers published in journals and conferences. His main research interests include resonant and soft-switching power converters, power converter topologies and control, grid-connected converters and system for renewable energy, modeling and stability of power converters, and envelop tracking power supply.

Prof. Ruan was a recipient of Sustainable Energy Systems Technical Achievement Award from IEEE Power Electronics Society in 2022, the Delta Scholarship by the Delta Environment and Education Fund in 2003 and was a recipient of the Special Appointed Professor of the Chang Jiang Scholars Program by the Ministry of Education, China, in 2007. From 2005 to 2013, and since 2017 again, he serves as Vice President of the China Power Supply Society (CPSS). From 2014 to 2016, he served as a Vice Chair of the Technical Committee on Renewable Energy Systems within the IEEE Industrial Electronics Society. Currently, he serves as an Editor for *IEEE Journal of Emerging and Selected Topics on Power Electronics* and an Associate Editor for *IEEE Transactions on Industrial Electronics, IEEE Transactions on Power Electronics, IEEE Transactions on Circuits and Systems–II*. He was the General Chair of IPEMC-ECCE Asia 2020 and the General Secretary of IPEMC-ECCE Asia 2009, a Technical Program Committee Chair of the IEEE 7th Annual Energy Conversion Congress and Exposition, and a Tutorial Committee Chair of the IEEE 12th Annual Energy Conversion Congress and Exposition.

Lihong Xie received the B.S. and Ph.D. degrees in electrical engineering from Nanjing University of Aeronautics and Astronautics (NUAA), Nanjing, China, in 2012 and 2018, respectively.

From November 2018 to August 2022, he worked as a research associate in the Electrical Energy Management Group (EEMG) at the University of Bristol as part of its work with the EPSRC Centre for Power Electronics. Since October 2022, he joined the College of Automation in NUAA as a Research Fellow. His current research interests include transformer modeling, conducted EMI and virtual prototyping of power converters.

Qing Ji received the B.S. and Ph.D. degrees in electrical engineering from Nanjing University of Aeronautics and Astronautics (NUAA), Nanjing, China, in 2006 and 2014, respectively.

Since December 2014, he joined the College of Mechanical and Electrical Engineering in Soochow University, and he is currently an Associate Professor. His research interests include power electronics and the mitigation of conducted EMI in power converters.

Xibo Yuan received the B.S. degree from the China University of Mining and Technology, Xuzhou, China, and the Ph.D. degree from Tsinghua University, Beijing, China, in 2005 and 2010, respectively, both in electrical engineering.

He has been a Professor since 2017 in the Electrical Energy Management Group, Department of Electrical and Electronic Engineering, University of Bristol, Bristol, UK. He also holds the Royal Academy of Engineering/Safran Chair in Advanced Aircraft Power Generation Systems. He is the Director of the UK National Centre for Power Electronics and an executive committee member of the IET Power Electronics, Machines and Drives (PEMD) network.

His research interests include power electronics and motor drives, wind power generation, multilevel converters, application of wide-bandgap devices, electric vehicles and more electric aircraft technologies. Professor Yuan is an Associate Editor of *IEEE Transactions on Industry Applications* and *IEEE Journal of Emerging and Selected Topics in Power Electronics*. He is a Fellow of IET and received The Isao Takahashi Power Electronics Award in 2018.

Chapter 1
Introduction

1.1 A Brief Introduction to the EMC and EMI

1.1.1 The Concept of EMC

Over the past few decades, there has been rapid development in microelectronics and power electronics technologies, significantly enhancing the quality of life and work for human society. Figure 1.1 illustrates the diverse applications of electrical and electronic equipments, including their use in electrified transportation, renewable energy generation, aerospace, information technology (IT), home appliance, and more.

However, the operation of these equipments produces electromagnetic emissions as a byproduct, which can propagate through conduction or radiation. These emissions have the potential to disrupt the normal operation of the power grid, degrade the performance of other equipments, interfere with telecommunications and data transfer, or even pose a threat to human safety by affecting navigation system. As the electromagnetic environment becomes increasingly complex, ensuring electromagnetic compatibility (EMC) of electrical and electronic equipments has become paramount. This is crucial to guarantee the smooth operation of equipment and prevent electromagnetic pollution in the environment.

EMC refers to the condition in which the electrical and electronic equipment can operate normally within its environment, and does not introduce electromagnetic emission that other nearby equipment can not tolerate [1, 2]. Consequently, EMC encompasses two aspects: electromagnetic interference (EMI) and electromagnetic susceptibility (EMS). EMI refers to the occurance where electromagnetic emissions interfere with the proper functioning of other equipments. On the other hand, EMS characterizes the ability of equipment to endure or resist the impact of electromagnetic emissions without experiencing malfunction.

© The Author(s), under exclusive license to Springer Nature Singapore Pte Ltd. 2024
X. Ruan et al., *Conducted Electromagnetic Interference in Power Converters: Modeling, Prediction and Reduction*, CPSS Power Electronics Series,
https://doi.org/10.1007/978-981-99-9295-9_1

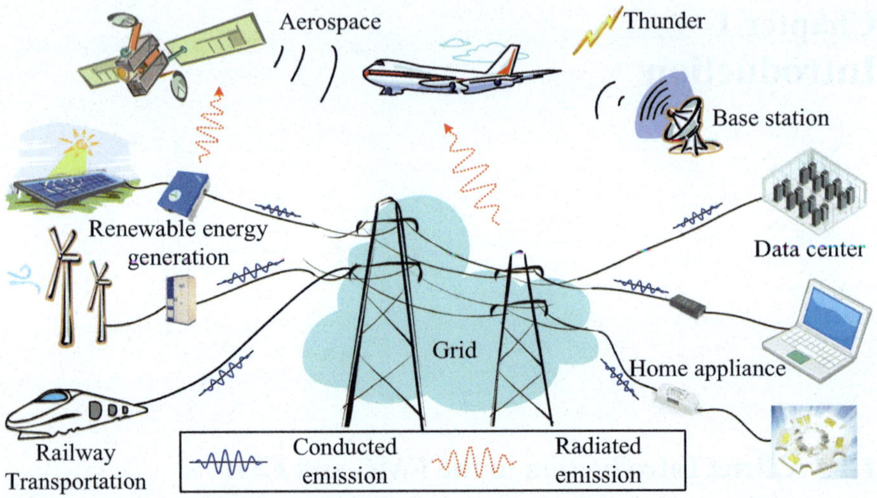

Fig. 1.1 The complex electromagnetic environment for electrical and electronic equipments

To address emerging EMI issues, the International Electrotechnical Commission (IEC) recommended the establishment of the "Comité International Spécial des Perturbations Radioélectriques" (CISPR) in Paris back in 1933. Since its inception, CISPR has played a pivotal role in developing various standards regarding the measurement and limits of EMI. For instance, CISPR 16 [3] delineates the equipment and procedure for EMI testing, while CISPR 22 [4] governs the measurement and limits for EMI from IT equipments. Today, the EMI standards established by CISPR have gained acceptance in most countries. In addition to CISPR, other organizations such as the Federal Communications Commission (FCC) and the Commission Européenne de Normalisation Électrotechnique (CENELEC) are actively involved in specifying related EMC standards. Examples include FCC Part 18 (for industrial, scientific and medical equipments) and EN 55025 (for boats and internal combustion engines). It is worth noting that EN 55022 [5] is functionally equivalent to CISPR 22.

1.1.2 The Concept of Conducted and Radiated EMI

Any EMI phenomenon comprises three essential elements, as shown in Fig. 1.2: noise source, coupling path and victim (or receiver). The noise source originates from natural phenomena or electrical and electronic equipments. When their electromagnetic emissions reach a certain level of strength, they create electromagnetic hazards. These hazards can lead to reduced performance, failures in other equipments, or even pose risks to the well-being of humans and other living organisms. The victim is the equipment that is susceptible to degradation or failure when exposed

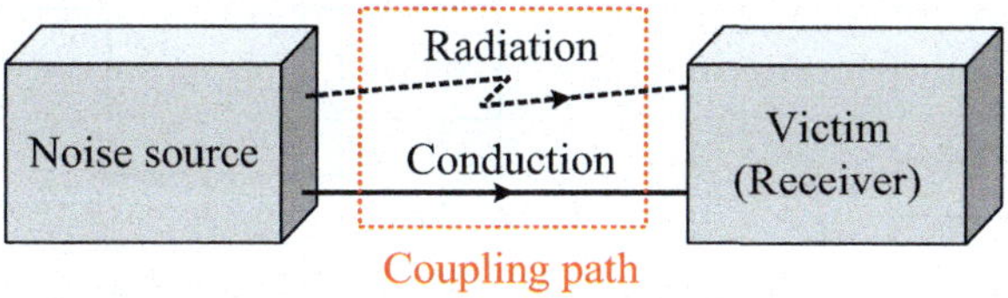

Fig. 1.2 Three elements of EMI

to electromagnetic emissions. The coupling path serves as a connection between the noise source and the victim, and it has two typical mechanisms: conduction and radiation. The conduction coupling path is created by a physical wire that connects the noise source and the victim, and the radiation coupling path extends through the surrounding space via electromagnetic radiation. Both conducted EMI and radiated EMI represent instances where electromagnetic emissions interfere with other equipment or even causes damage, and their coupling paths are conduction and radiation, respectively.

Figure 1.3a, b illustrate typical conduction coupling paths. The direct conduction coupling occurs when two circuits are physically connected by wires. The common-impedance coupling happens when the currents of two circuits share a common impedance. This is commonly observed in power distribution and grounding systems. Figure 1.3c–e show the radiation coupling paths, which can be classified into near-field and far-field coupling. The electric coupling and magnetic coupling belong to near-field coupling, which can be analyzed by a circuit-based approach. The far-field coupling is usually analyzed by a field-based approach, as it is the electromagnetic wave propagation between two circuits.

Conducted emission generated by the equipment can enter the power grid via direct conduction coupling. Consequently, conducted EMI standards normally stipulate limits for conducted emissions through the input power cord of the equipment. For example, the CISPR 22 standard defines Quasi Peak (QP) and Average (AV) limits for IT equipments in Class A and Class B, as shown in Fig. 1.4. Equipments in Class A are intended for use in commercial and industrial environment, while Class B equipments are designed for residential use.

When initially formulating EMC standards, CISPR's primary goal was to prevent interference with radio communication and broadcast signals. In civil applications, long-distance communication and broadcasting typically occur within the frequency range of [150 kHz, 1 GHz], since the electromagnetic waves below 150 kHz cannot propagate through the Earth's ionosphere. Besides, conducted currents above 30 MHz experience significantly attenuation, owing to the impedance and associated parasitic inductance of the power cord. Therefore, CISPR's stipulated frequency range for conducted EMI is [150 kHz, 30 MHz]. Further details on the measurement of conducted EMI and the principle of the related instruments will be provided in Sect. 2.1 of Chap. 2.

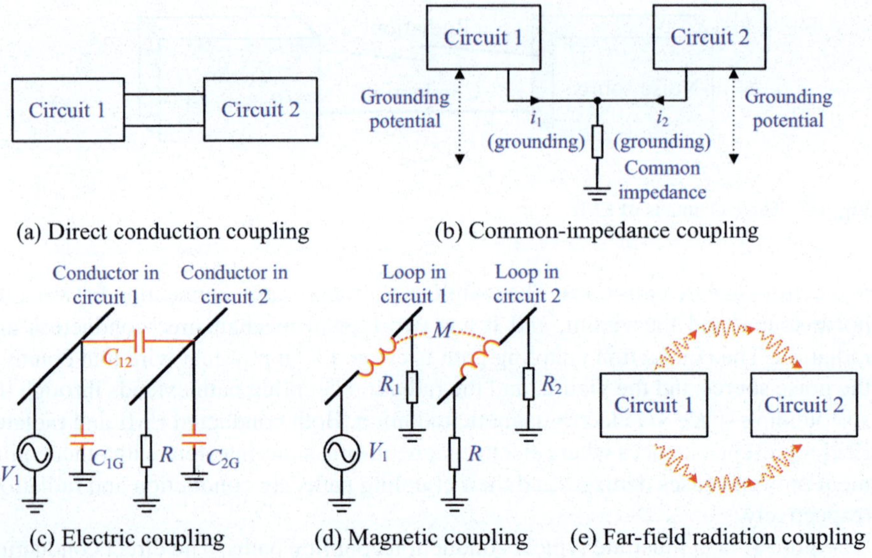

Fig. 1.3 Typical coupling paths of conduction and radiation

Fig. 1.4 Conducted EMI limits of EN 55022 for equipments in Class A and B

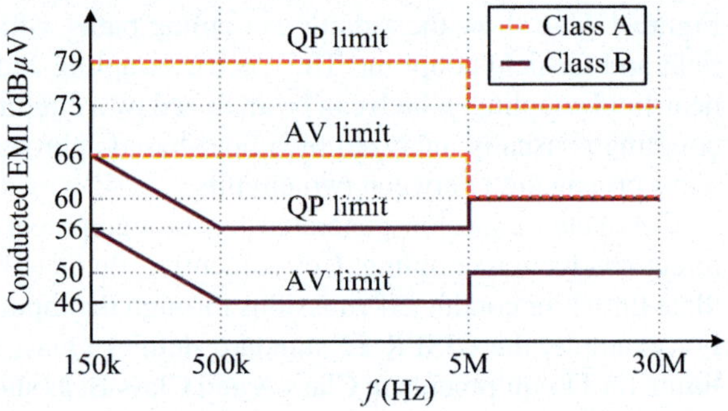

1.2 Conducted EMI in Power Converters

With the rapid advancement of power electronics technologies, power converters have found widespread applications across various sectors, including general industry, power system, electrified transportation, IT technology, aerospace, home appliance [6], and etc. Power devices in power converters are operated in high-frequency switching mannar, resulting in high rates of change in voltage (dv/dt) and current (di/dt) in the circuit. As a result, conducted emissions are generated in the input power cord, either through the input wire or near-field coupling. This emission not only pollutes the grid, but also disrupts the operation of surrounding equipment, presenting significant challenges related to conducted EMI [7–9].

Essentially, the conducted EMI appears as the noise currents in the input power cord, which can be separated into common-mode (CM) and differential-mode (DM) ones. Figure 1.5 shows the propagation path for the CM and DM currents in a power converter with an AC input. The CM noise current i_{CM} is the component that has the same magnitude and direction in the input currents, which propagates through both the input cords and returns to the protective earth (PE). The DM noise current i_{DM} is the component that has the same magnitude but opposite direction in the input currents, and it propagates within the input power cords and forms a closed circuit via the input source.

In general, the DM noise in a power converter is associated with the input current caused by switching actions, while the CM noise is generated by the displacement current flowing through parasitic capacitance between high dv/dt nodes in the circuit and the PE [10–15]. For non-isolated dc-dc converters, consider the boost converter shown in Fig. 1.6a as an example. During switching actions, the electric potential of the drain of switch Q_b undergoes abrupt changes, leading to the generation of displacement current. This displacement current flows through the parasitic capacitance C_p, untimately entering the PE and causing CM noise. C_p primarily consists of the parasitic capacitance between the drain of Q_b and the PE via the heatsink as shown in Fig. 1.6b, along with the stray capacitance between the trace of printed circuit board (PCB) and the PE.

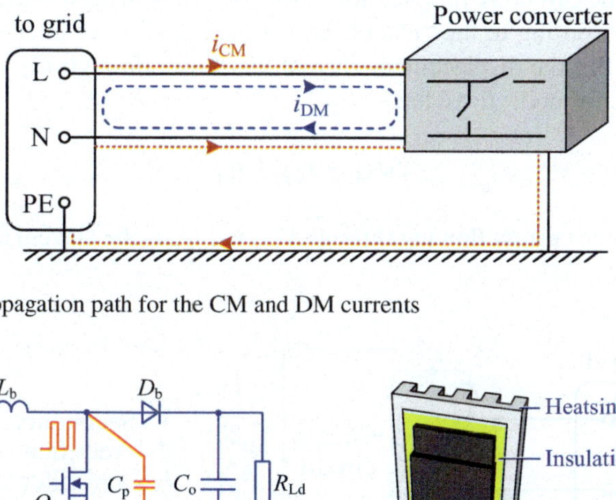

Fig. 1.5 The propagation path for the CM and DM currents

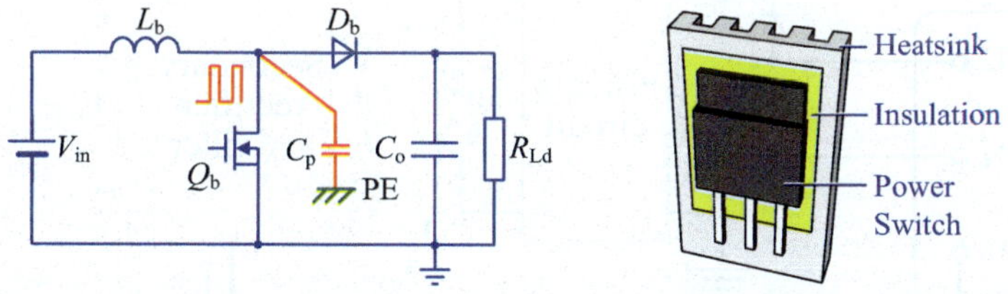

(a) Boost converter (b) The origin of parasiric capacitance C_p

Fig. 1.6 The CM noise propagation path of boost converter

In high-power applications, safety considerations often lead to the heatsink connected to the PE. In such cases, the dominant propagation path for the CM noise is through the parasitic capacitance C_p between the drain of the power switch and the heatsink [16]. C_p is associated with the thickness and permittivity of the insulation as well as the relative area of the power switch and the heatsink. Typically, C_p falls within the range of tens of picofarads.

Conversely, in low-power applications with plastic enclosure, the heatsink is often connected to the primary ground (PG). Consequently, the displacement current between the power switch and the heatsink remains within the circuit, which does not contribute to the CM noise. In this case, the dominant CM noise propagation path becomes the stray capacitance between the PCB trace and the PE. However, the CM noise is relatively low due to the small value of stray capacitance.

Figure 1.7 shows the CM noise propagation path in an isolated power converter with AC input. The isolated power converter generally consists of input rectifier bridge, primary circuit, transformer, and secondary rectifier and filter. PG denotes the primary ground, and SG represents the secondary ground. In this converter, C_p represents the parasitic capacitance between the high dv/dt nodes in the primary circuit and the PE, C_s stands for the parasitic capacitance between the high dv/dt nodes in the secondary rectifier and filter and the PE, and C_{ps} is the inter-winding capacitance of the transformer. In cases where a three-wire input is used, SG is usually connected to PE for safety reasons.

In Fig. 1.7, the CM noise propagation path is drawn with the dashed line, including C_p and C_{ps}. According to the cutset between the transformer inter-winding capacitance and the PE, the displacement current that flows through C_{ps} is equal to i_{Cps}. Applying KCL for node J, we have

$$2i_{CM} = i_{Cp} + i_{Cps} \tag{1.1}$$

where, $2i_{CM}$ is the current flowing through PE, and i_{Cp} is the current flowing through C_p.

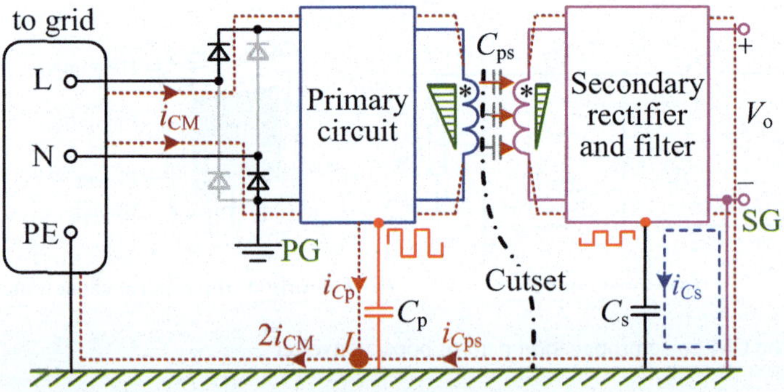

Fig. 1.7 The CM noise propagation path of a general isolated dc-dc converter

As noted in (1.1), both i_{Cp} and i_{Cps} contribute to the CM noise current i_{CM}. Therefore, C_p and C_{ps} are the critical components for propagating the CM noise in isolated power converters [17]. Note that, when SG is connected to PE, the displacement current i_{Cs} does not contribute to the CM noise since it does not flow back to the input side.

In power converters, the CM and DM noise each have distinct origins, and consequently, require different noise reduction methods. Therefore, separating the CM and DM noises is essential for several purposes, including diagnosing the conducted EMI spectrum, selecting the conducted EMI reduction methods, and designing the EMI filter. In Sect. 2.2 of Chap. 2, the separation of CM and DM noises and the design of EMI filter will be presented. The remainder of this book will cover the modeling, prediction and reduction of the conducted EMI in ac-dc rectifiers, dc-dc converters, and dc-ac inverters. As the ac-ac cycloconverter is relatively infrequently applied, its conducted EMI is not addressed in this monograph.

1.3 Modeling and Prediction of the CM and DM EMI in AC-DC Rectifiers

The ac-dc rectifier serves to transform an ac voltage into a dc voltage, and power factor correction (PFC) is usually incorporated for realizing unity power factor (PF). The PFC converter shapes the input current to a sinusoidal waveform that is in-phase with the input voltage, and also regulates the output voltage.

There are various topologies for PFC converters, with the boost PFC converter being the most widely used, as shown in Fig. 1.8. This popularity is attributed to its relatively low ripple in the input current due to the presence of the input inductor L_b, a high PF, and a simple circuit configuration.

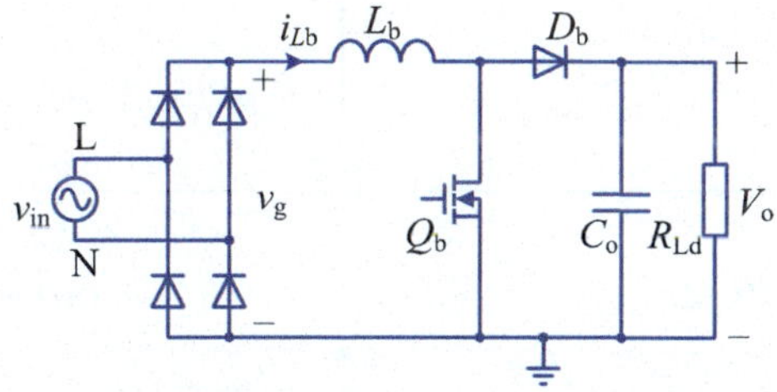

Fig. 1.8 The topology of boost PFC converter

1.3.1 Operating Modes and Control Schemes for Boost PFC Converter

The boost PFC converter has three typical operating modes, i.e., continuous conduction mode (CCM), critical conduction mode (CRM) and discontinuous conduction mode (DCM). Figure 1.9 shows the waveform of the boost inductor current i_{Lb} in these operating modes.

Under CCM, the inductor current ripple is very small, and high PF can be achieved. As a result, CCM is preferred for medium-to-high power applications. Figure 1.10 illustrates the average-current-control diagram for the CCM boost PFC converter [18]. In this diagram, the converter operates using pulse-width modulated, and the switching frequency is constant. The output voltage is sensed and compared with the voltage reference V_{o_ref}, The error signal in this comparison is sent into the voltage regulator $G_v(s)$. The output of $G_v(s)$ and the sensed input rectified voltage (referred to as "waveform input") are multiplied, serving as the current reference i_{Lb_ref} for the inner current loop. The sensed inductor current is then compared with i_{Lb_ref}, and the error signal is sent into the current regulator $G_i(s)$. The output of $G_i(s)$ is compared with a sawtooth carrier signal, resulting in the generation of the driving signals for the switch Q_b.

Under CRM, the boost diode is naturally turned off without experiencing reverse recovery, the power switch realizes zero-current-switching (ZCS), and high PF is achieved. Therefore, CRM is preferred in low-to-medium power conversions [19–21]. Figure 1.11 shows the control diagram of CRM boost PFC converter. In this diagram, the output of the voltage regulator $G_v(s)$ and the sensed input rectified voltage are multiplied, serving as the peak reference for the inductor current, denoted as i_{Lb}. When i_{Lb} drops to zero, the zero-crossing detector (ZCD) is triggered, turning on the switch Q_b. When i_{Lb} touches the peak reference, Q_b is turned off. Under CRM,

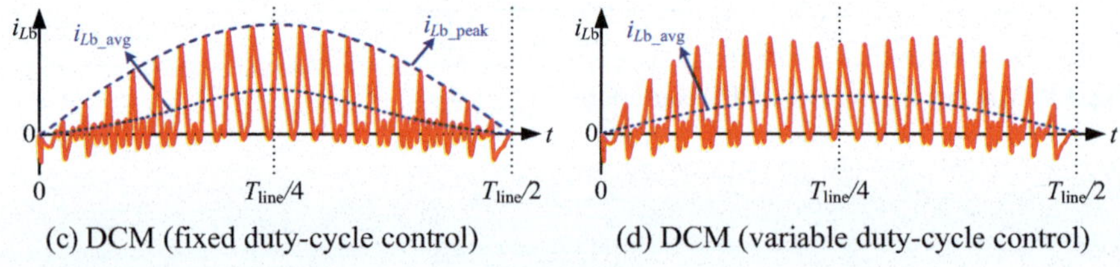

Fig. 1.9 Operation modes of boost PFC converter

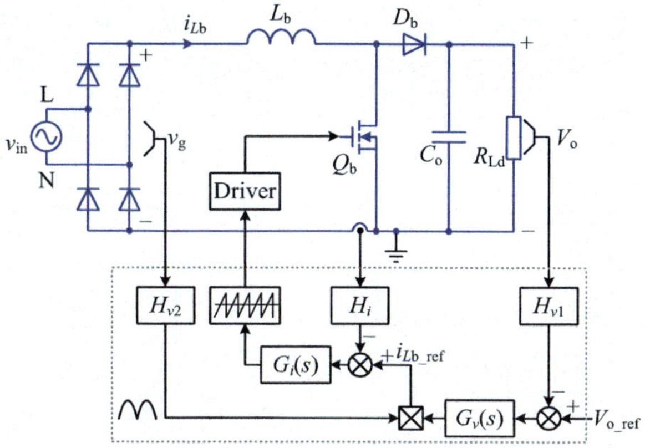

Fig. 1.10 The CCM boost PFC with average current control

i_{Lb} follows a triangular waveform within a switching cycle, with its average value being half of the peak current. As the peak reference for i_{Lb} tracks a sinusoidal pattern, the average value of i_{Lb} also becomes sinusoidal, leading to the realization of PFC. Under CRM, the switching frequency varies in a half-line cycle, and it is influenced by the input voltage and load. As a result, the conducted EMI characteristic of CRM boost PFC converter is quite complex.

The DCM boost PFC converter offers certain advantages such as ZCS for the power switch and no reverse recovery for the diode. There are two primary control schemes for DCM boost PFC converters: fixed duty-cycle control [22] and variable duty-cycle control [23], as shown in Fig. 1.12a, b, respectively.

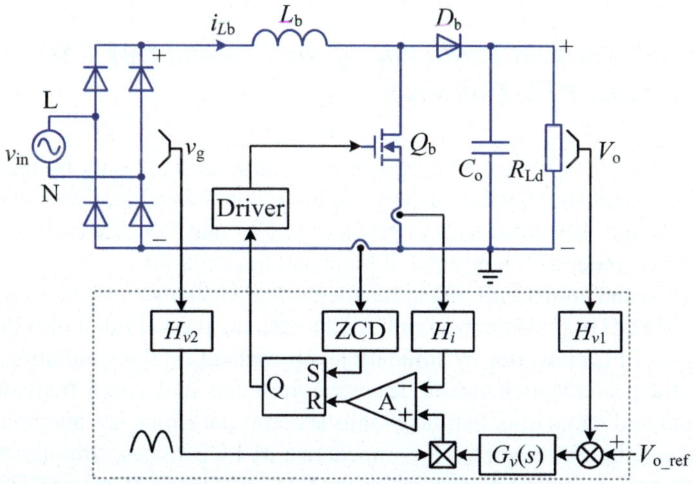

Fig. 1.11 The CRM boost PFC converter

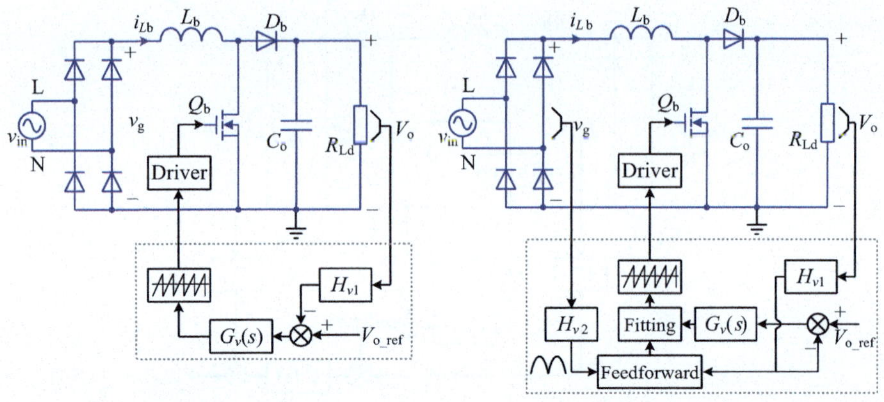

(a) Fixed duty-cycle control (b) Variable duty-cycle control

Fig. 1.12 The DCM boost PFC converter

As shown in Fig. 1.12a, the control circuit of the DCM boost PFC converter with the fixed duty-cycle control is relatively simple. However, the average value of the inductor current, as depicted in Fig. 1.9c, is not a sine wave in a half-line cycle, leading to a low PF. To achieve a high PF, the variable duty-cycle control is employed, as shown in Fig. 1.12b. The variable duty-cycle control is achieved by calculating the duty cycle of the converter that leads to unity PF, then utilizing input voltage feedforward control and fitting algorithms to generate the desired duty cycle.

In fact, the CCM boost PFC converter will partially operate in DCM under light load condition, and DCM will occur in half-line cycle as the load becomes lighter.

1.3.2 Modeling and Prediction of the CM and DM EMI in Boost PFC Converter

The conducted EMI of boost PFC converter is influenced not only by the operating modes and control strategy, but also by the input voltage and load. For assisting the EMI filter design, it is necessary to predict the CM and DM EMI spectra of boost PFC converter under different input voltage and load condition.

Simulation and numerical calculation methods are usually employed to predict the CM and DM EMI of power converters. In [24–28], the prediction of the CM and DM EMI in boost PFC converter by simulation is presented. In the simulation, practical models of the power switch and diode are incorporated, and a high-frequency model of inductors and capacitors that takes into account parasitics are adopted. Furthermore, the simulation integrates the impedance of PCB traces. Simulation method proves effective in accurately predicting the CM and DM EMI spectra. However, to determine the worst-case CM and DM EMI spectra for the EMI filter design, it is necessary to simulate and compare CM (DM) EMI under various input voltage and load conditions, which is quite time-intensive. Moreover, the distinct CM and DM EMI spectra characteristics of the power converter under different input voltage and load condition remain unclear.

The numerical calculation method begins by identifying the CM and DM noise propagation paths of the converter. Subsequently, and the CM and DM noise equivalent circuits are established, respectively. Then, calculations are performed to determine the spectrum of the noise source, CM noise and DM noise. Finally, the CM and DM EMI spectra of the converter are predicted based on the operation principles of the EMI receiver. This method mainly considers the main circuit parameters and the dominant CM parasitic capacitance, making it suitable for predicting the CM and DM EMI at low frequencies. In [29], the numerical calculation method is employed to predict the CM and DM EMI in a CRM boost PFC converter. This approach successfully identifies the worst-case CM and DM EMI spectra to assist the EMI filter design. However, this method does not provide a detailed analytical understanding of the CM and DM EMI of the converter.

In Chap. 3, the CM and DM noise equivalent circuit of boost PFC converter will be firstly derived. Then, the influence of the main circuit parameters on the EMI filter performance will be discussed, and the EMI filter topology suitable for boost PFC converter will be presented. Based on the CM and DM noise equivalent circuits, the conducted EMI spectra characteristics of boost PFC converter with different control schemes will be further analyzed in Chaps. 4 and 5. In Chap. 4, the conducted EMI of average-current-controlled boost PFC converter will be studied, and the relationship between the operation modes and the input voltage and load conditions will be revealed. With the analysis of the CM and DM noise spectra of the converter under different operation modes, the worst case of the CM and DM EMI spectra are predicted, and the EMI filter design is given. In Chap. 5, the conducted EMI of CRM boost PFC converter is investigated. By analyzing the conducted EMI spectra characteristics under different input voltage and load conditions, the worst-case CM and DM EMI spectra are predicted to assist the EMI filter design.

1.4 Modeling and Reduction of the CM EMI in DC-DC Converters

1.4.1 Modeling of the CM EMI in DC-DC Converters

DC-DC converters are used to generate a regulated DC voltage from an input DC voltage, which can be categorized into non-isolated and isolated ones. The DM EMI of DC-DC converters is relatively easy to be analyzed and reduced, since its propagation path is between the input power cords. Comparatively, managing the CM noise is challenging, owing to its complicated propagation path.

In non-isolated DC-DC converters, the dominant CM noise propagation path is the parasitic capacitance between the high dv/dt nodes and the PE. Deriving its CM noise equivalent circuit is relatively straightforward. However, in isolated DC-DC converters, the inter-winding capacitance of the high-frequency transformer is another critical propagation path for the CM noise. To achieve high efficiency, transformer windings are tightly wound to reduce the leakage inductance and AC resistance of the windings, leading to a substantial inter-winding capacitance between the primary and secondary windings. In wire-wound transformers, the inter-winding capacitance usually ranges from tens of pF to hundreds of pF. For planar transformers, due to their large relative area and small distance between the windings, the inter-winding capacitance is even larger, ranging from hundreds of pF to potentially thousands of pF. The displacement current flowing through the inter-winding capacitance is related to not only the capacitance, but also the electric potential distribution within the transformer windings. Consequently, the CM noise issue in isolated DC-DC converters is more complex than that in non-isolated ones.

Since the inter-winding capacitance presents complexities and inconvenient for circuit analysis, it is usually equivalently converted into lumped capacitances by keeping the displacement current unchanged. In [30], the inter-winding capacitance of the transformer in a flyback converter is represented by two lumped capacitances, C_{ac} and C_{bd}, as shown in Fig. 1.13a. In [31], the transformer inter-winding capacitance in a dual-switch forward converter is replaced by two lumped capacitances, C_{ac} and C_{bc}, as shown in Fig. 1.13b.

Basically, in the existing literatures, the lumped capacitances are determine by calculating the displacement current for a specific transformer winding structure, and the derived lumped capacitance may not be feasible for other topologies. As a solution to this issue, Chap. 6 will present a generalized lumped capacitance model for modeling the CM noise of a general isolated DC-DC converter with general transformer winding structure. Based on this model, the equivalent noise source (ENS) will be proposed for analyzing the CM noise of isolated DC-DC converters.

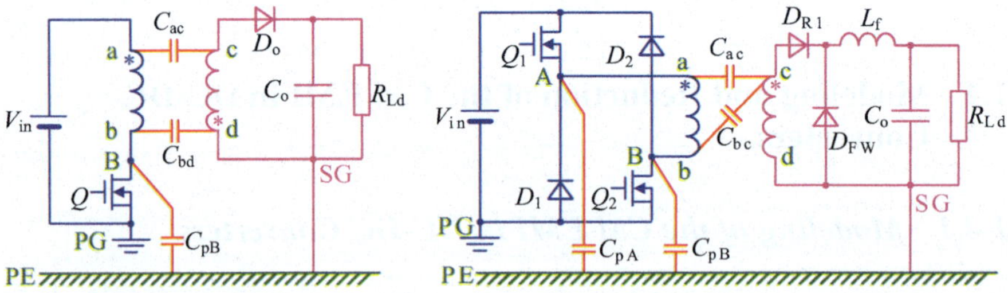

(a) Flyback converter (b) Dual-switch forward converter

Fig. 1.13 The lumped capacitance model in flyback and dual-switch forward converters

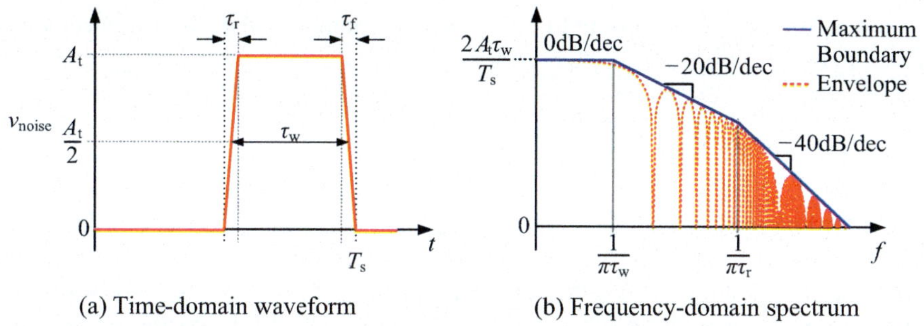

Fig. 1.14 The noise voltage with typical trapezoidal waveform in the time-domain and the envelope, maximum boundary in the frequency-domain

1.4.2 Reduction of the CM EMI in DC-DC Converters

For DC-DC converters, the CM EMI filters usually occupy significant volume and weight, leading to reduced power density and efficiency of the converter. Therefore, it is necessary to suppress the original CM noise of DC-DC converters, in order to reduce the EMI filter size. Various techniques have been proposed to reduce the CM noise for DC-DC converters, such as the suppression methods at the source side, shielding technique and winding cancellation method in the transformer, symmetry circuits, the passive cancellation methods, and etc.

1. Conducted EMI Reduction at the Source Side

In general, the voltage across the main switches in DC-DC converters can be treated as the noise source that generates the conducted EMI. Figure 1.14a shows a typical noise voltage v_{noise} with a trapezoidal waveform, where, A_t is the amplitude, T_s is the switching period, τ_w is the duration when the voltage is rising and falling at A_t / 2, τ_r and τ_f are the rising and falling time, respectively. Here, τ_r and τ_f is assumed to be the same.

The harmonic amplitude of a trapezoidal waveform can be expressed as [32]

$$V_{noise\_amp}(kf_s) = 2A_t\frac{\tau_w}{T_s}\cdot\left|\frac{\sin(\pi\tau_w kf_s)}{\pi\tau_w kf_s}\right|\cdot\left|\frac{\sin(\pi\tau_r kf_s)}{\pi\tau_r kf_s}\right| \tag{1.2}$$

where, k is the harmonic order, and f_s is the switching frequency.

In (1.2), the harmonic amplitude only appears at discrete frequencies kf_s. To better show the distribution of harmonic amplitude, the envelope can be obtained by replacing kf_s in (1.2) with a continuous frequency f. As a result, the envelope $V_{noise\_env}(f)$ of the spectrum of a trapezoidal waveform is derived as

$$V_{noise\_env}(f) = 2A_t\frac{\tau_w}{T_s}\cdot\left|\frac{\sin(\pi\tau_w f)}{\pi\tau_w f}\right|\cdot\left|\frac{\sin(\pi\tau_r f)}{\pi\tau_r f}\right| \tag{1.3}$$

Since $|\sin(x)| \leq 1$, according to (1.3), the maximum boundary for the envelope is given by

$$V_{\text{noise\_mb}}(f) = 2A_t \frac{\tau_w}{T_s} \cdot \left(\frac{1}{\pi \tau_w f}\right) \cdot \left(\frac{1}{\pi \tau_r f}\right) \tag{1.4}$$

According to (1.3) and (1.4), the envelope and maximum boundary of the trapezoidal waveform are depicted, as shown in Fig. 1.14b. As seen, the maximum boundary has a slope of 0 dB/dec at the frequencies below $1/(\pi \tau_w)$, − 20 dB/dec at the frequencies within $[1/(\pi \tau_w), 1/(\pi \tau_r)]$, and − 40 dB/dec at the frequencies above $1/(\pi \tau_r)$. By increasing τ_r, the maximum boundary of the spectrum will be reduced for the frequencies above $1/(\pi \tau_r)$. Soft-switching technique helps reduce the conducted EMI at high frequencies since the rising/falling time τ_r is increased at the switching transients [33–35].

2. Shielding Technique in Transformers

The transformer shielding technique can be single- and double-shielding. Figure 1.15a shows the single-shielding technique, where the shielding layer S_{P1} is inserted between the primary winding W_{P1} and secondary windings W_{S1}, and it is connected to the PG of the converter [36]. Although the electric coupling between W_{P1} and W_{S1} is blocked, the electric coupling between S_{P1} and its adjacent windings emerges. This coupling gives rises to the displacement current and consequently lead to the generation of the CM noise.

Figure 1.15b shows the double-shielding technique, where two shielding layers that have the same winding direction between the primary and secondary winding are inserted. The one close to the primary winding is connected to the PG and the other one close to the secondary winding is connected to the SG. With such arrangement, the two shielding layers have identical electric potential distribution, and thus there is no displacement current between them [37]. However, for the interleaved winding structure, each pair of double shielding layers should be inserted between every two adjacent primary and secondary windings, which will occupy large window area and increase the complexity of manufacture.

3. Winding Cancellation Method in Transformers

The winding cancellation method is designed to effectively eliminate the displacement current between the adjacent primary and secondary windings in a transformer. Figure 1.16 shows the CM noise circuit of the flyback converter and its transformer structure that adopts the winding cancellation method [38]. In this transformer, the W_{P1}, W_{P2} and W_{P3} layers of the primary winding use enameled wire, while the W_{P4} layer of the primary winding uses copper foil with proper height to cover the whole window height. With this setup, the electric coupling between W_{P3} and the secondary winding W_{S1} is blocked. Besides, the electric coupling between W_{P3} and W_{S1} can also be neglected, since it is blocked by the tightly wound layers W_{P2} and W_{P4}. Importantly, the turns of W_{P4} is the same as that of W_{S1}, and the dotted terminals B and C on the primary and secondary sides are equi-potential in the conducted EMI

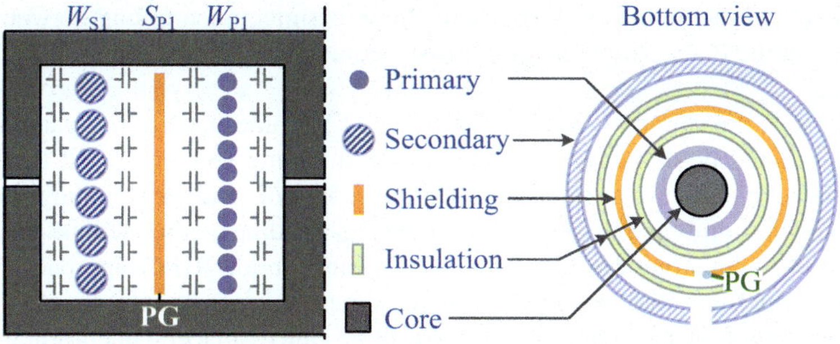

(a) The transformer with single shielding layer

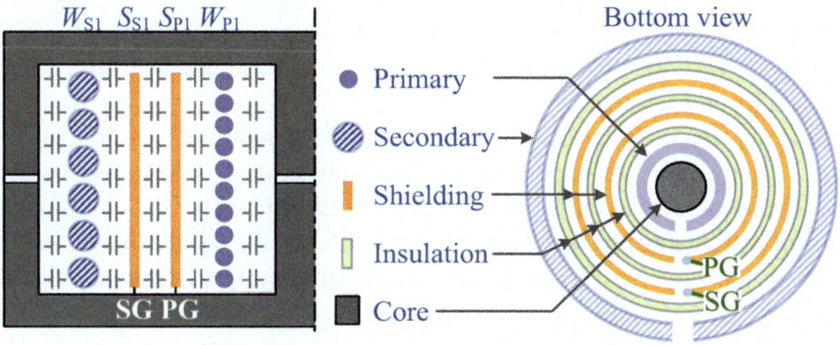

(b) The transformer with double shielding layers

Fig. 1.15 The transformer with shielding layers

frequency range. As a result, W_{P4} and W_{S1} have the same electric potential distribution [38], leading to the elimination of the displacement current between W_{P4} and W_{S1}, and thus the CM noise is reduced.

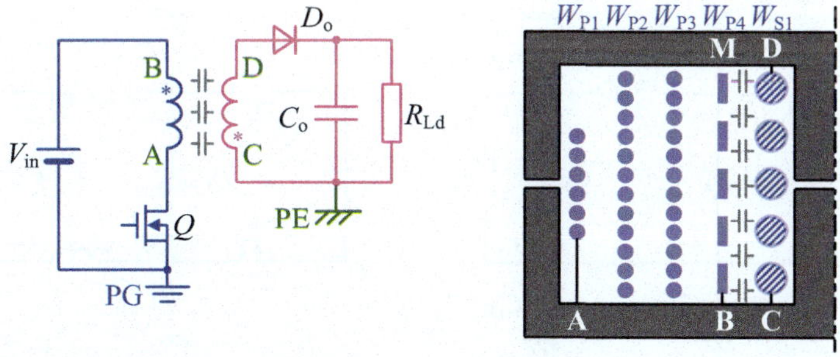

Fig. 1.16 The flyback converter and the transformer with winding cancellation

However, as shown in Fig. 1.16, there still exists electric coupling between W_{S1} and W_{P1} via the core, which generates displacement current and contributes to the CM noise.

In Chap. 7, the shielding-based approach will be introduced to mitigate the CM noise in isolated dc-dc converters. The single-shielding technique will be further analyzed to explore ways to enhance CM noise reduction, and the shielding-winding method and the combined shielding and balancing-winding method will be proposed. Furthermore, a shielding-based hybrid passive cancellation (HPC) approach will be proposed, which not only cancels the displacement current flowing through transformer inter-winding capacitance, but also nullifies the displacement current flowing through the parasitic capacitance between the high dv/dt nodes in the primary circuit and the PE.

4. Symmetry Circuit

Figure 1.17 shows both the asymmetrical and symmetrical buck converters [39] along with their CM parasitic capacitances. As shown in Fig. 1.17a, in the asymmetrical buck converter, the high dv/dt switching node A generates displacement current that flows through the parasitic capacitance C_{pA}, resulting in the CM noise. On the other hand, in the symmetrical buck converter shown in Fig. 1.17b, the electric potential waveforms with respect to the negative terminal O of the input voltage are given in Fig. 1.18. When switches Q_1 and Q_2 are conducting, the electric potentials of nodes A and B are $v_{AO} = V_{in}$ and $v_{BO} = 0$, and the voltage applied across inductors L_{f1} and L_{f2} is both $(V_{in} - V_o)/2$. Consequently, the electric potentials of the nodes C and D are

$$\begin{cases} v_{CO} = v_{AO} - v_{Lf1} = (V_{in} + V_o)/2 \\ v_{DO} = v_{CO} - V_o = (V_{in} - V_o)/2 \end{cases} \quad (1.5)$$

When Q_1 and Q_2 are turned off, the freewheeling diode D_{FW} is conducting. Thus, Q_1 and Q_2 sustain a voltage of $V_{in}/2$, and $v_{AO} = v_{BO} = V_{in}/2$. As the voltage across the inductor L_{f1} is $-0.5V_o$, we have $v_{CO} = (V_{in} + V_o)/2$ and $v_{DO} = (V_{in} - V_o)/2$.

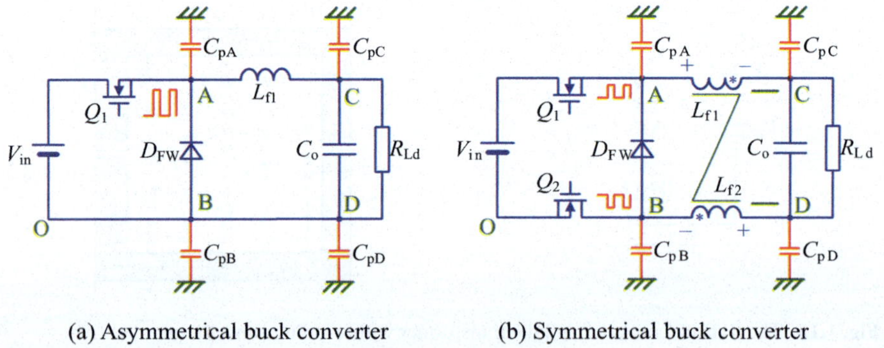

Fig. 1.17 The asymmetrical and symmetrical buck converter

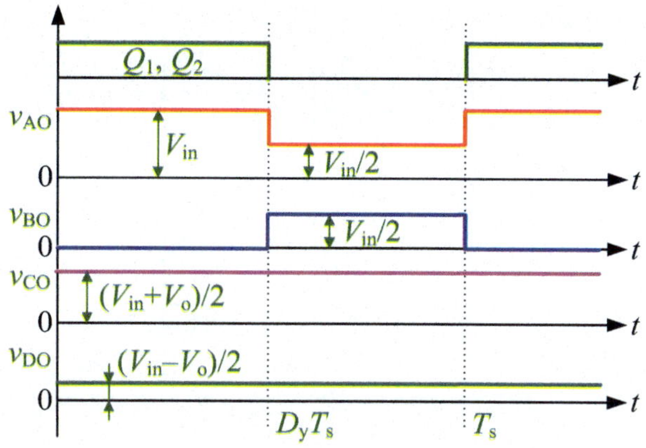

Fig. 1.18 The electric potential waveforms of the symmetrical buck converter

Obviously, v_{CO} and v_{DO} only have dc component, and there is no displacement current flowing through C_{pC} and C_{pD}. Besides, v_{AO} and v_{BO} have high dv/dt characteristics and are complementary to each other. When $C_{pA} = C_{pB}$, the displacement current flowing through C_{pA} and C_{pB} can be effectively cancelled.

Compared to the asymmetrical circuit, the symmetrical circuit requires more components, leading to higher cost and power losses. However, for the circuit that is inherently symmetrical, the CM noise can be reduced with appropriate control scheme. For example, the primary circuit of dual-switch forward converter shown in Fig. 1.13b is already symmetrical. Since Q_1 and Q_2 are switched simultaneously, the electric potential of nodes A and B are complementary. By equalizing the capacitance C_{pA} and C_{pB}, the resultant displacement current can be cancelled by each other. It is worth noting that the secondary rectifier and filter are not symmetrical, and the ratio of the lumped capacitances C_{ac} and C_{bc} should satisfy a certain value determined by the transformer turns ratio. Thus, the CM noise can be eliminated [31].

Figure 1.19 shows the full-bridge converter, which is inherently a symmetrical topology. However, when phase-shift control is implemented, the electric potentials of the two phase-leg midpoints change at different time, and the generated displacement current flowing through the associated parasitic capacitance, including the parasitic capacitance between the phase-leg midpoints and the PE, as well as the transformer inter-winding capacitance can not be cancelled by each other [40, 41]. Furthermore, the voltage across the resonant inductor affects the electric potentials of the transformer primary winding terminals, which also generates the CM noise. In Chap. 8, the CM noise of the phase-shifted full-bridge (PSFB) converter will be investigated, and the CM noise reduction method will be proposed.

5. Passive Cancellation Method

The passive cancellation method is implemented by introducing a compensation branch that generates a compensation current aimed at cancelling the CM noise current. In non-isolated (isolated) DC-DC converters, the voltage across the inductor (transformer winding) usually contains the same ac component as the drain-to-source

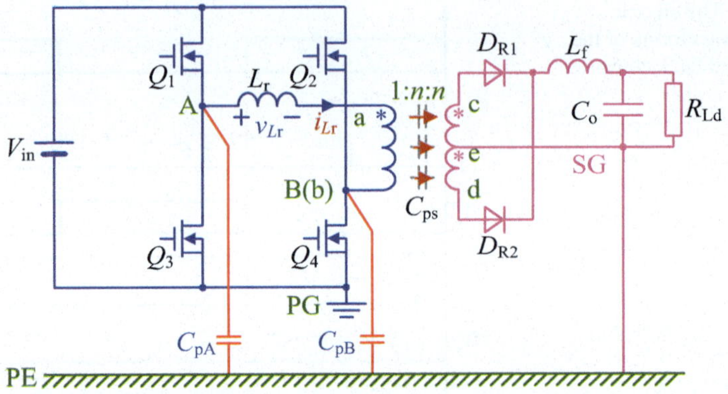

Fig. 1.19 The PSFB converter and its CM noise propagation path

voltage of the switches. Therefore, a compensation winding can be coupled with the inductor (transformer) to induce a compensation voltage that is out-of-phase with the d*v*/d*t* in the main circuit. By properly selecting the compensation capacitance, the compensation current will cancel the CM noise current effectively.

Figure 1.20 shows the applications of passive cancellation method in buck converter and flyback converter [42]. As seen, the compensation voltage across the compensation winding is out-of-phase with respect to the high d*v*/d*t* node in the circuit, as long as the coupling coefficient is close to unity.

However, it is important to note that the performance of CM noise cancellation at high frequencies may be compromised or deteriorate, owing to the parasitics of the cancellation branch. For instance, the AC resistance of the compensation winding will restrict the magnitude of compensation current at high frequencies, leading to degradation in the CM noise reduction performance [42]. Additionally, due to the presence of leakage inductance and the parasitic capacitance of the compensation winding, the compensation voltage and the node with high d*v*/d*t* in the converter are no longer out-of-phase at high frequencies, ultimately resulting in poorer CM noise [42].

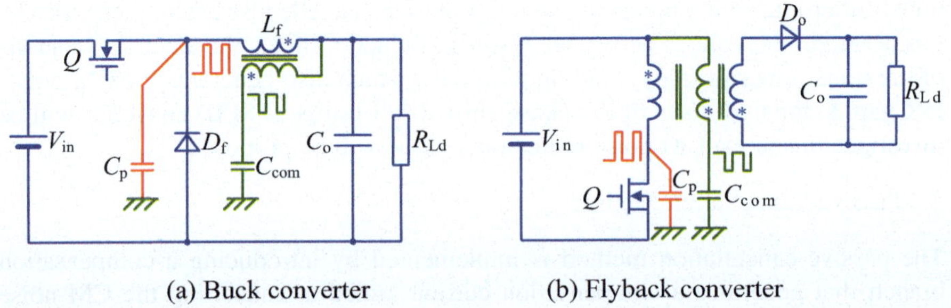

Fig. 1.20 The Buck and flyback converter adopting passive cancellation method

From the description above, it becomes evident that the underlying principles of CM noise reduction by symmetry circuit and passive cancellation method are essentially the same. Both methods aim to eliminate the CM noise current by introducing a cancellation branch in parallel with the converter for providing the compensation current. In Chap. 9, instead of cancelling the CM current, the CM voltage cancellation (CMVC) method will be explored, which introduces a compensation voltage in series with the converter to reduce the CM voltage. In Chap. 10, based on the CMVC method, the split-winding (SW) configuration for non-isolated DC-DC converters will be introduced, allowing for simultaneous cancellation of the CM currents at the input and output sides.

1.5 Modeling and Reduction of the CM EMI in DC-AC Inverters

The DC-AC inverter is employed to convert a DC voltage to an AC voltage, and is widely applied in the applications such as motor drive, induction heating, uninterruptable power supply and renewable energy generation. Normally, the input and output of the inverter do not share the same ground, giving rise to the CM voltage and subsequent CM current. This CM current flows between the CM impedance at the input and output sides, leading to the conducted EMI and leakage current in the DC-AC inverter systems.

In photovoltaic (PV) inverter systems, non-isolated inverters are usually adopted to achieve high efficiency and small volume/weight. However, due to the presence of the parasitic capacitance C_{PV} between the PV panel and the PE, the CM voltage of the inverter will generate the CM current through C_{PV}, the parasitic capacitance between the inverter and the PE, the output filter, line impedance Z_{line} and the grounding impedance Z_g, as shown in Fig. 1.21a [43, 44]. The CM current at the output side will flow into the grid and cause conducted EMI, while the CM current at the input side may cause safety hazard.

In motor drive systems [45], as shown in Fig. 1.21b, the input CM current goes to the DC source and may introduce the conducted EMI. Meanwhile, the output CM current, often referred to as leakage current, will flow into the motor bearings and potentially cause bearing failures. When the fast-switching silicon-carbide (SiC) and gallium-nitride (GaN) devices are incorporated, the dv/dt in the circuit is significantly higher than that in silicon device-based circuits, exacerbating the CM noise issues. Therefore, reducing the CM current at the input and output sides in inverter systems is a critical concern [46–49].

Methods of mitigating the CM EMI and leakage current in inverter systems fall into three main categories: the filter approach [50–53], the CM voltage reduction methods, and the CM noise cancellation methods. Passive filters attenuate the CM noise by incorporating the CM inductors along the CM noise propagation path to increase the impedance, and by adding Y capacitors to bypass the noise into the PE.

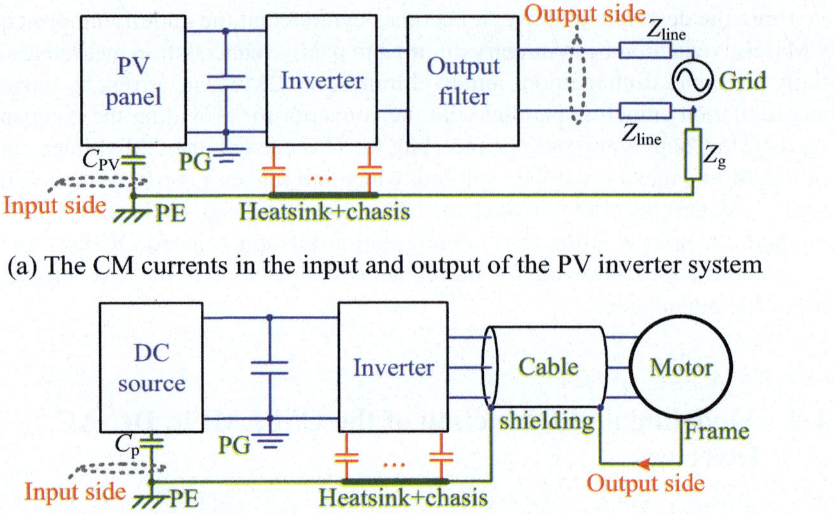

(a) The CM currents in the input and output of the PV inverter system

(b) The CM currents in the input and output of motor drive system

Fig. 1.21 Two typical inverter systems

It is important to note that, as the filter design for both sides is coupled due to the CM loop, the attenuation of CM noise at one side could make the CM noise at the other side even worse, since the Y capacitors alter the impedance of the CM loop [53]. To address this issue, a floating EMI filter is proposed in [53], which connects the AC-side Y capacitors to the midpoint of the DC-link, allowing for independent filter design on each side. On the other hand, active filters employ active devices to sense and compensate the CM noise [45]. When designing the passive or active filter, the CM impedance of the source and the load must be considered, since they affect the CM noise attenuation of the filter.

Basically, the CM voltage of the inverter can be reduced either by modifying the inverter topology or by applying advanced modulation strategy, and thus the CM noise can be suppressed. It is revealed that the inverter topology with constant CM voltage does not generate the CM noise. For the PV inverter systems, various topologies with constant CM voltage have been proposed [54–56]. For motor drive systems, relevant topologies can be found in [57–59]. However, the increased hardware cost of these topologies should be concerned in practical applications. Alternatively, the advanced modulation strategy aims to modify the CM voltage spectra and reduce its harmonic amplitude, hence attenuating the CM noise [60–63]. It is important to recognize that reducing the CM noise reduction through advanced modulation may involve trade-offs with other factors, such as switching losses, output voltage harmonics and neutral point voltage balance [64].

To effectively cancel the CM noise, the CM transformer (CMT) is normally incorporated to cancel the CM voltage of DC-AC inverters. The CMT consists of tightly coupled sensing winding and injecting windings, where, the sensing winding captures the CM voltage of the inverter, while the injecting windings induce the compensation voltage and injects it into the output of the inverter [65–69]. Thus, the CM voltage of the inverter is cancelled, and the CM current at the output side is attenuated as a result. However, addressing CM noise reduction at the input side remains a challenge.

In Chap. 11, a comprehensive CM noise model of inverter systems will be presented, which takes into account the input and output CM impedance. Based on the CMVC method in Chap. 9, the cancellation method will be proposed. This method involves inserting two CMTs into the inverter input and output sides, in order to reduce the CM current at the input and output sides of the inverter simultaneously. The influence of the parasitics in the CMT will be investigated as they limit the bandwidth for CM noise cancellation.

1.6 Summary

With the widespread use of power electronics technology, the issues posed by conducted EMI in power converters have grown increasingly significant. In this chapter, the basic concept of the EMC and conducted EMI is briefly introduced, and the generation of the CM and DM EMI in power converters is explained. The research status and key issues for the conducted EMI in power converters are elaborated, including the modeling, prediction and reduction of the conducted EMI in AC-DC rectifiers, DC-DC converters, and DC-AC inverters.

References

1. D. Weston, *Electromagnetic Compatibility: Principles and Applications*, 2nd edn. (Marcel Dekker Inc., New York, 2001)
2. F. Costa, C. Gautier, E. Laboure, B. Revol, *Electromagnetic Compatibility in Power Electronics* (ISTE Ltd., London, 2014)
3. CISPR 16-1-2, *Specification for Radio Disturbance and Immunity Measuring Apparatus and Methods—Part 1–2* (British Standards, 2009)
4. CISPR 22, *Information Technology Equipment Radio Disturbance Characteristics Limits and Methods of Measurement* (International Electrotechnical Commission, 2005)
5. EN 55022, *Limits and Methods of Measurement of Radio Disturbance Characteristics of Information Technology Equipment* (European Norm Standard, 2006)
6. X. Ruan, *Power Electronics* (China Machine Press, Beijing, 2021) (in Chinese)
7. W. Ma, L. Zhang, J. Meng, *EMC in the Independent Power Systems and the Power Electronics Equipments* (Science Press, Beijing, 2007) (in Chinese)
8. W.O. Henry, *Electromagnetic Compatibility Engineering* (Wiley, New York, 2009)
9. K. Mainali, R. Oruganti, Conducted EMI mitigation techniques for switch-mode power converters: a survey. IEEE Trans. Power Electron. **25**(9), 2344–2356 (2010)
10. X. Wu, N.-.K Poon, C.-M. Lee, B.M.-H. Pong, Z.-M. Qian, A study of common mode noise in switching power supply from a current balancing viewpoint, in *Proceedings of IEEE International Conference on Power Electronics and Driving Systems* (1999), pp. 621–625

11. P. Kong, S. Wang, F.C. Lee, Improving balance technique for high frequency common mode noise reduction in boost PFC converters, in *Proceedings of IEEE Power Electronics Specialists Conference (PESC)* (2008), pp. 2941–2947
12. P. Kong, S. Wang, F.C. Lee, Common mode EMI noise suppression for bridgeless PFC converters. IEEE Trans. Power Electron. **23**(1), 291–297 (2008)
13. Z. Lu, W. Chen, Common mode EMI noise reduction technique by noise path configuration of high frequency power transformer, in *Proceedings of IEEE International Power Electronics and Motion Control Conference (IPEMC-ECCE Asia)* (2009), pp. 954–956
14. P. Kong, Y. Jiang, F.C. Lee, Common mode EMI noise characteristics of low-power ac–dc converters. IEEE Trans. Power Electron. **27**(2), 731–738 (2012)
15. M. Pahlevaninezhad, D. Hamza, P.K. Jain, An improved layout strategy for common-mode EMI suppression applicable to high-frequency planar transformers in high-power dc/dc converters used for electric vehicles. IEEE Trans. Power Electron. **29**(3), 1211–1228 (2014)
16. S. Wang, P. Kong, F.C. Lee, Common mode noise reduction for boost converters using general balance technique. IEEE Trans. Power Electron. **22**(4), 1410–1416 (2007)
17. Y. Chu, S. Wang, A generalized common mode current cancellation approach for power converters. IEEE Trans. Industr. Electron. **62**(7), 4130–4140 (2015)
18. C. Silva, *UC3854 Controlled Power Factor Correction Circuit Design*. Unitrode Corp, Application Note U-134 (1999)
19. J.-S. Lai, D. Chen, Design consideration for power factor correction boost converter operating at the boundary of continuous conduction mode and discontinuous conduction mode, in *Proceedings of IEEE Applied Power Electronics Conference and Exposition (APEC)* (1993), pp. 267–273
20. J. Zhang, J. Shao, P. Xu, F.C. Lee, M.M. Jovanovic, Evaluation of input current in the critical mode boost PFC converter for distributed power systems, in *Proceedings of IEEE Applied Power Electronics Conference and Exposition (APEC)* (2001), pp. 130–136
21. M. Marvi, A. Fotowat-Ahmady, A fully ZVS critical conduction mode boost PFC. IEEE Trans. Power Electron. **27**(4), 1958–1965 (2012)
22. K. Liu, Y. Lin, Current waveform distortion in power factor correction circuits employing discontinuous-mode boost converters, in *Proceedings of IEEE Power Electronics Specialists Conference (PESC)* (1989), pp. 825–829
23. K. Yao, X. Ruan, X. Mao, Z. Ye, Variable-duty-cycle control to achieve high input power factor for DCM boost PFC converter. IEEE Trans. Industr. Electron. **58**(5), 1856–1865 (2011)
24. L. Rossetto, S. Buso, G. Spiazzi, Conducted EMI issue in a 600-W single-phase boost PFC design. IEEE Trans. Ind. Appl. **36**(2), 578–585 (2000)
25. E. Hertz, Thermal and EMI modeling and analysis of a boost PFC circuit designed using a genetic-based optimization algorithm. Master dissertation, Virginia Polytechnic Institute and State University, Blacksburg, USA, 2001
26. L. Yang, Modeling and characterization of a PFC converter in the medium and high frequency ranges for predicting the conducted EMI. Master dissertation, Virginia Polytechnic Institute and State University, Blacksburg, USA, 2003
27. L. Yang, B. Lu, W. Dong, Z. Lu, M. Xu, F.C. Lee, W.G. Odendaal, Modeling and characterization of a 1 kW CCM PFC converter for conducted EMI prediction, in *Proceedings of IEEE Applied Power Electronics Conference and Exposition (APEC)* (2004), pp. 763–769
28. F. Giezendanner, J. Biela, J.W. Kolar, S. Zudrell-Koch, EMI noise prediction for electronic ballasts. IEEE Trans. Power Electron. **25**(8), 2133–2141 (2010)
29. Z. Wang, S. Wang, P. Kong, F.C. Lee, DM EMI noise prediction for constant on-time, critical mode power factor correction converters. IEEE Trans. Power Electron. **27**(7), 3150–3157 (2012)
30. J. He, W. Chen, J. Jiang, Model and analysis on common-mode conducted EMI of a switched mode power supply, in *Proceedings of the Chinese Society for Electrical Engineering (CSEE)*, vol. 25, no. 8 (2005), pp. 50–55 (in Chinese)
31. P. Kong, S. Wang, F.C. Lee, Reducing common-mode noise in two-switch forward converter. IEEE Trans. Power Electron. **26**(5), 1522–1533 (2011)

32. C.R. Paul, *Introduction to Electromagnetic Compatibility*, 2nd edn. (Wiley, New York, 2006)
33. P. Caldeira, R. Liu, D. Dalal, W. Gu, Comparison of EMI performance of PWM and resonant power converters, in *Proceedings of IEEE Power Electronics Specialists Conference (PESC)* (1993), pp. 134–140
34. M. Joshi, V. Agarwal, Design optimization of ZVS and ZCS quasi-resonant converters for EMI reduction, in *Proceedings of International Conference on Electromagnetic Interference and Compatibility (ICEIC)* (1997), pp. 407–413
35. H. Chung, S.-Y. Hui, K.-K. Tse, Reduction of power converter EMI emission using soft-switching technique. IEEE Trans. Electromagn. Compat. **40**(3), 282–287 (1998)
36. Y. Yang, D. Huang, F.C. Lee, Q. Li, Analysis and reduction of common mode EMI noise for resonant converters, in *Proceedings of IEEE Applied Power Electronics Conference and Exposition (APEC)* (2014), pp. 566–571
37. Y. Yang, D. Huang, F.C. Lee, Q. Li, Transformer shielding technique for common mode noise reduction in isolated converters, in *Proceedings of IEEE Energy Conversion Congress and Exposition (ECCE)* (2013), pp. 4149–4153
38. P. Kong, F.C. Lee, Transformer structure and its effects on common mode EMI noise in isolated power converters, in *Proceedings of IEEE Applied Power Electronics Conference and Exposition (APEC)* (2010), pp. 1424–1429
39. M. Shoyama, G. Li, T. Ninomiya, Balanced switching converter to reduce common-mode conducted noise. IEEE Trans. Industr. Electron. **50**(6), 1095–1099 (2003)
40. P. Kong, Common mode EMI noise reduction techniques in switch mode power supplies. Ph.D. Dissertation, Virginia Polytechnic Institute and State University, Blacksburg, USA, 2009
41. I.A. Makda, M. Nymand, Common-mode noise analysis, modeling and filter design for a phase-shifted full-bridge forward converter, in *Proceedings of IEEE International Conference on Power Electronics and Drive Systems (PEDS)* (2015), pp. 1100–1105
42. D. Cochrane, D. Chen, D. Boroyevic, Passive cancellation of common-mode noise in power electronic circuits. IEEE Trans. Power Electron. **18**(3), 756–763 (2003)
43. W. Li, Y. Gu, H. Luo, W. Cui, X. He, C. Xia, Topology review and derivation methodology of single-phase transformerless photovoltaic inverters for leakage current suppression. IEEE Trans. Industr. Electron. **62**(7), 4537–4551 (2015)
44. H. Xiao, S. Xie, Leakage current analytical model and application in single-phase transformer-less photovoltaic grid-connected inverter. IEEE Trans. Electromagn. Compat. **52**(4), 902–913 (2010)
45. S. Wang, Y.Y. Maillet, F. Wang, D. Broyevich, R. Burgos, Investigation of hybrid EMI filters for common-mode EMI suppression in a motor drive system. IEEE Trans. Power Electron. **25**(4), 1034–1045 (2010)
46. S. Chen, T.A. Lipo, D. Fitzgerald, Modeling of motor bearing currents in PWM inverter drives. IEEE Trans. Ind. Appl. **32**(4), 1365–1370 (1996)
47. N. Mutoh, M. Nakanishi, M. Kanesaki, J. Nakashima, EMI noise control methods suitable for electric vehicle drive systems. IEEE Trans. Electromagn. Compat. **47**(4), 930–937 (2005)
48. F. Wang, Motor shaft voltages and bearing currents and their reduction in multilevel medium-voltage PWM voltage-source-inverter drive applications. IEEE Trans. Ind. Appl. **36**(5), 1336–1341 (2000)
49. J. Adabi, F. Zare, G. Leawich, A. Ghosh, Leakage current and common mode voltage issues in modern AC drive systems, in *Proceedings of Australaisian University Power Engineering Conference* (2007), pp. 1–6
50. H. Akagi, T. Shimizu, Attenuation of conducted EMI emissions from an inverter-driven motor. IEEE Trans. Power Electron. **23**(1), 282–290 (2008)
51. D. Xu, Q. Gao, W. Wang, Design of a passive filter to reduce common-mode and differential-mode voltage generated by voltage-source PWM inverter, in *Proceedings of Annual Conference of IEEE Industrial Electronics Society (IECON)* (2006), pp. 2483–2487
52. D. Dong, X. Zhang, F. Luo, D. Boroyevich, P. Mattavelli, Common-mode EMI noise reduction for grid-interface converter in low-voltage dc distribution system, in *Proceedings of IEEE Applied Power Electronics Conference and Exposition (APEC)* (2012), pp. 451–457

53. Y. Liu, Z. Mei, S. Jiang, W. Liang, Conducted common-mode electromagnetic interference suppression in the AC and DC sides of a grid-connected inverter. IET Power Electron. **13**(13), 2926–2934 (2020)
54. D. Han, C.T. Morris, B. Sarlioglu, Common-mode voltage cancellation in PWM motor drives with balanced inverter topology. IEEE Trans. Industr. Electron. **64**(4), 2683–2688 (2017)
55. C.T. Morris, D. Han, B. Sarlioglu, Reduction of common mode voltage and conducted EMI through three-phase inverter topology. IEEE Trans. Power Electon. **32**(3), 1720–1724 (2017)
56. S. Heribert, S. Christoph, K. Jurgen, Inverter for transforming a dc voltage into an ac current or an ac voltage. Europe Patent 1 369 985 (A2) (2003)
57. R. Gonzalez, J. Lopez, P. Sanchis, L. Marroyo, Transformerless inverter for single-phase photovoltaic systems. IEEE Trans. Power Electron. **22**(2), 693–697 (2007)
58. B. Yang, W. Li, Y. Gu, W. Cui, X. He, Improved transformerless inverter with common-mode leakage current elimination for a photovoltaic grid-connected power system. IEEE Trans. Power Electron. **27**(2), 752–762 (2012)
59. M. Victor, F. Greizer, S. Bremicker, U. Hubler, Method of converting a direct current voltage from a source of direct current voltage, more specifically from a photovoltaic source of direct current voltage, into a alternating current voltage. U.S. Patent 7 411 802 B2 (2008)
60. X. Yuan, J. Yon, P. Mellor, Common-mode voltage reduction in three-level neutral-point-clamped converters with neutral point voltage balance. in *Proceedings of IEEE International Symposium on Industrial Electronics* (2013), pp. 1–6
61. Y. Han, H. Lu, Y. Li, J. Chai, Analysis and suppression of shaft voltage in SiC-based inverter for electric vehicle applications. IEEE Trans. Power Electron. **34**(7), 6276–6285 (2019)
62. H. Zhang, A.V. Jouanne, S. Dai, A.K. Wallace, F. Wang, Multilevel inverter modulation schemes to eliminate common-mode voltages. IEEE Trans. Ind. Appl. **36**(6), 1645–1653 (2000)
63. M. Cacciato, A. Consoli, G. Scarcella, A. Testa, Reduction of common-mode currents in PWM inverter motor drives. IEEE Trans. Ind. Appl. **35**(2), 469–476 (1999)
64. D. Jiang, Z. Shen, Q. Li, J. Chen, Z. Liu, *Advanced Pulse-Width-Modulation: With Freedom to Optimize Power Electronics Converters* (Springer, 2021)
65. S. Ogasawara, H. Ayano, H. Akagi, An active circuit for cancellation of common-mode voltage generated by a PWM inverter. IEEE Trans. Power Electron. **13**(5), 835–841 (1998)
66. M.M. Swamy, K. Yamada, T. Kume, Common mode current attenuation techniques for use with PWM drives. IEEE Trans. Power Electron. **16**(2), 248–255 (2001)
67. Y. Murai, T. Kubota, Y. Kawase, Leakage current reduction for a high-frequency carrier inverter feeding an induction motor. IEEE Trans. Ind. Appl. **28**(4), 858–863 (1992)
68. M.C.D. Piazza, M. Luna, G. Vitale, EMI reduction in dc-fed electric drives by active common-mode compensator. IEEE Trans. Electromagn. Compat. **56**(5), 1067–1076 (2014)
69. Y. Jiang, D. Xu, X. Chen, A novel inverter output dv/dt suppression filter, in *Proceedings of Annual Conference of IEEE Industrial Electronics Society (IECON)* (2003), pp. 2901–2905

Chapter 2
Measurement of Conducted Electro-magnetic Interference (EMI) and Design of EMI Filter

2.1 Measurement of Conducted Electromagnetic Interference (EMI)

2.1.1 *Test Setup for Measuring the Conducted EMI*

Figure 2.1 shows the test setup for measuring the conducted EMI within the frequency range [150 kHz, 30 MHz], where, a line impedance stabilization network (LISN) is added between the equipment under test (EUT) and the input source. The LISN provides a specific impedance to ensure the repeatability of the test results, and converts the noise current generated by the EUT into the noise voltage for the convenience of measurement. The noise voltage from the LISN is sent into the EMI receiver for acquiring the conducted EMI spectrum of the EUT.

Figure 2.2 shows the arrangement of the instruments for measuring the conducted EMI stipulated in CISPR 22. The EUT is placed on a wooden table with a height of 0.8 m, and keeps a horizontal distance of 0.4 m from the vertical reference plane. The LISN is tightly grounded to the horizontal reference plane, and keeps a horizontal distance of 0.8 m from the EUT. The minimal dimension for the vertical and horizontal reference planes is 2 m × 2 m.

2.1.2 *Line Impedance Stabilization Network*

As seen from Fig. 2.1, the LISN is a multiport passive network composed of capacitors, inductors and resistors, and it has the selective properties for signals at different frequencies. According to CISPR-16-1 [1], when measuring the conducted EMI, the parameters in the LISN are: $C_1 = 1\ \mu$F, $L_1 = 50\ \mu$H, $C_2 = 0.1\ \mu$F, $R_1 = 1$ kΩ. At the line frequency, e.g., 50 Hz, the impedance of inductor L_1 is very small and

© The Author(s), under exclusive license to Springer Nature Singapore Pte Ltd. 2024

X. Ruan et al., *Conducted Electromagnetic Interference in Power Converters: Modeling, Prediction and Reduction*, CPSS Power Electronics Series,
https://doi.org/10.1007/978-981-99-9295-9_2

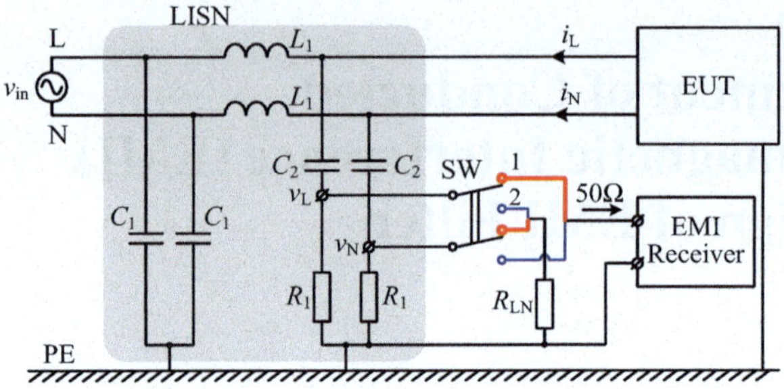

Fig. 2.1 The test setup for measuring the conducted EMI

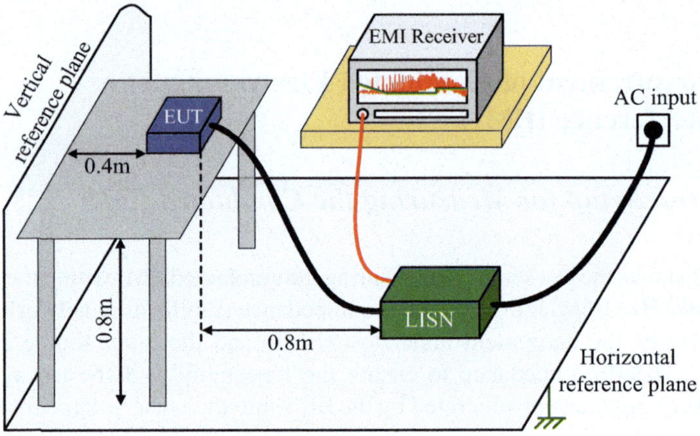

Fig. 2.2 The arrangement of the instruments for measuring the conducted EMI

has negilible effect on power delivery. At the conducted EMI frequency range from 150 kHz to 30 MHz, the impedance of C_1 is very small, and the noise at the input side is bypassed by C_1 without affecting the test results. The noise currents i_L and i_N generated by the EUT mainly flow through the branch consisted of capacitor C_2, test resistor R_{LN} (with $R_{LN} = 50\ \Omega$, significantly smaller than R_1), and the input side of the EMI receiver (which also has 50-Ω input resistance [1]). As a result, the noise voltages, v_L and v_N, are generated [2]. When measuring v_L from the Live line, switch SW is placed at position 1, and the test terminal of v_N in the LISN is connected to R_{LN}. When measuring v_N from the Neutral line, switch SW is placed at position 2, and the test terminal of v_L in the LISN is connected to R_{LN}. Additionally, when the LISN is disconnected from the power lines, R_1 acts as the bleeding resistors for the capacitors in the LISN to prevent electric shock.

2.1.3 Separation of the CM and DM Noises

The common-mode (CM) and differential-mode (DM) EMI have distinct origins, and their respective suppression methods are different. Therefore, separation of the conducted EMI into CM and DM noises is convenient for diagnosing the conducted EMI spectrum, assisting the selection of suppression methods, and designing the EMI filters.

Figure 2.3 shows the block diagram of CM and DM noise separation. The noise separator [3] is designed with 50-Ω input impedance, so that the impedance from the LISN side for the conducted EMI is unchanged.

In Fig. 2.3, i_L and i_N are the noise currents from the L line and N line, respectively, and i_{CM} and i_{DM} are the CM and DM noise currents, respectively. So, we have

$$\begin{cases} i_L = i_{CM} + i_{DM} \\ i_N = i_{CM} - i_{DM} \end{cases} \tag{2.1}$$

According to (2.1), i_{CM} and i_{DM} can be calculated as

$$\begin{cases} i_{CM} = \frac{1}{2}(i_L + i_N) \\ i_{DM} = \frac{1}{2}(i_L - i_N) \end{cases} \tag{2.2}$$

In Fig. 2.3, v_L and v_N are the test voltages across the equivalent 50-Ω resistor generated by i_L and i_N, respectively, and we have $v_L = 50i_L$, $v_N = 50i_N$. The noise separator receives v_L and v_N as the input, and outputs the CM or DM voltages, v_{CM} or v_{DM}, expressed as

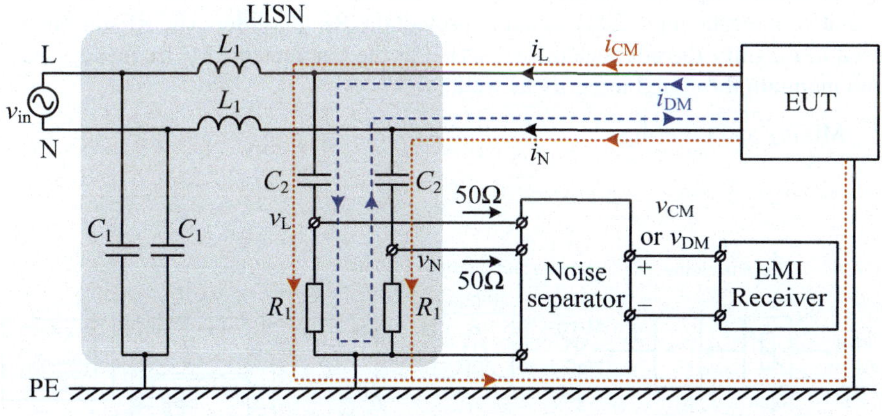

Fig. 2.3 The block diagram of CM and DM noise separation

$$\begin{cases} v_{CM} = \frac{1}{2}(v_L + v_N) = \frac{1}{2} \times 50 \cdot (i_L + i_N) \\ v_{DM} = \frac{1}{2}(v_L - v_N) = \frac{1}{2} \times 50 \cdot (i_L - i_N) \end{cases} \tag{2.3}$$

According to (2.2) and (2.3), we have

$$\begin{cases} v_{CM} = 50 i_{CM} \\ v_{DM} = 50 i_{DM} \end{cases} \tag{2.4}$$

As seen from (2.4), v_{CM} and v_{DM} are equivalently the voltages across the 50-Ω test resistor that carries i_{CM} and i_{DM}, respectively. In this sense, the noise separation is accomplished.

2.1.4 EMI Receiver

Figure 2.4 shows the basic structure of the EMI receiver [2], including the input attenuator, radio-frequency (RF) amplifier, frequency controller, mixer, oscillator, intermediate-frequency (IF) filter, envelope detector, and the detectors for acquiring Peak (PK), Quasi-Peak (QP) and Average (AV) values. The EMI signal can be either the noise voltages v_L and v_N according to the conducted EMI standard, or the CM and DM voltages v_{CM} and v_{DM} when noise separation is required. As shown in Fig. 2.4, the EMI signal is processed with three major steps, namely, pre-processing, mixing and filtering, and detection.

1. Pre-processing

In Fig. 2.4, the input attenuator provides 50-Ω input impedance for the EMI receiver, and attenuates the input EMI signal to protect the RF amplifier. The RF amplifier is used to magnify the attenuated EMI signal at the conducted EMI frequency range, and maintain the signal integrity as well.

2. Mixing and Filtering

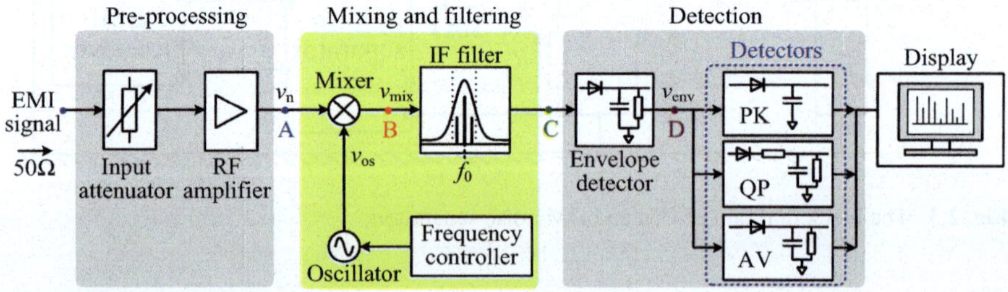

Fig. 2.4 The basic structure of the EMI receiver

Figure 2.5 gives the signal flow in the mixing and filtering step. Firstly, the mixer multiplies the EMI signal v_n with the signal v_{os} generated by the oscillator. v_{os} is a sine wave with the frequency of f_{os}, amplitude of 1 and initial phase angle of φ_{os}, expressed as

$$v_{os} = \sin(2\pi f_{os}t + \varphi_{os}) \tag{2.5}$$

For the convenience of illustration, v_n is also assumed to be a sine wave with a fixed frequency of f_e ($f_e \ll f_{os}$), amplitude of A_e and initial phase angle of φ_e, expressed as

$$v_n = A_e \sin(2\pi f_e t + \varphi_e) \tag{2.6}$$

The output signal of the mixer, v_{mix}, is the product of v_n and v_{os}. According to (2.5) and (2.6), we have

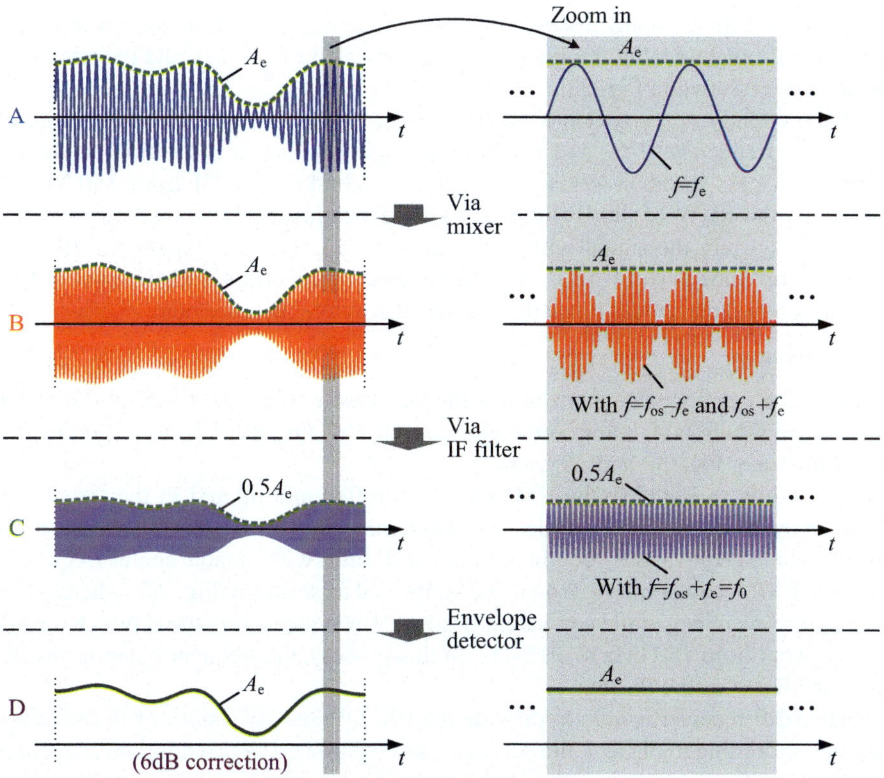

Fig. 2.5 The signal flow and processing steps in the EMI receiver

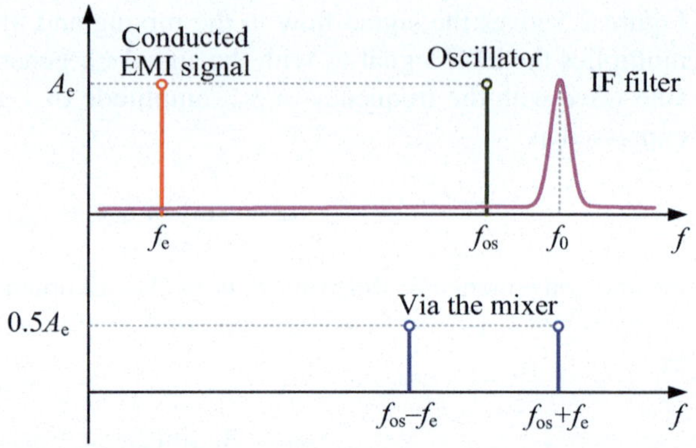

Fig. 2.6 The function of the mixer

$$v_{mix} = v_{os}v_n = 0.5A_e\left\{\begin{array}{c}\cos[2\pi(f_{os} - f_e)t + (\varphi_{os} - \varphi_e)] \\ + \cos[2\pi(f_{os} + f_e)t + (\varphi_{os} + \varphi_e)]\end{array}\right\} \quad (2.7)$$

As seen from (2.7), the mixer shifts the EMI signal with the frequency of f_e into the mixed signal with the frequency of $(f_{os} + f_e)$ and $(f_{os} - f_e)$, and the amplitude is halved, as shown in Fig. 2.6.

Subsequently, v_{mix} is sent into the IF filter. The IF filter is a band-pass filter centered at a fixed frequency of f_0, and it allows the signal within its resolution bandwidth (RBW) to pass through. When measuring the conducted EMI from 150 kHz to 30 MHz, the RBW of the IF filter is 9 kHz. By adjusting f_{os} to make $(f_{os} + f_e)$ equal to f_0, only the signal with the frequency of $(f_{os} + f_e)$ passes the IF filter, and its amplitude is still $0.5A_e$. As the amplitude of the original EMI signal is A_e, a correction of 6 dB should be made to achieve accurate measurement.

3. Detection

Finally, the envelope detector outputs the envelope of the signal from the IF filter, whose amplitude is A_e after correction, and the PK, QP, AV detectors acquires the corresponding PK, QP and AV values.

According to Fig. 2.6, the EMI signal is shifted and filtered in the frequency-domain by the frequency controller, oscillator, mixer and the IF filter. Therefore, these blocks can be equivalently combined into an IF filter with tunable center frequency, and the EMI receiver in Fig. 2.4 can be simplified as shown in Fig. 2.7, where, f_{IF} is the center frequency of the equivalent IF filter. When measuring the conducted EMI, f_{IF} is swept from 150 kHz to 30 MHz. In this process, the frequency step is smaller than the RBW of the IF filter.

The EMI receiver is able to provide the PK, QP and AV values of the envelope signal. According to Fig. 2.4, the detectors are constructed by resistors and capacitors with different time constants for charging and discharging processes, and the main features are:

(1) For the PK detector, the time constant for charing is zero, and the time constant for discharging is infinite. Therefore, the output capacitor voltage rises very

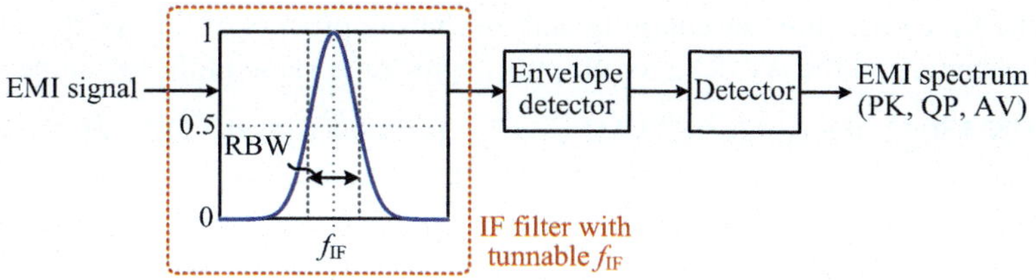

Fig. 2.7 The simplified structure of EMI receiver

quickly to follow the peak value of the envelope signal, and keeps unchanged when the envelope signal is decreasing. The capacitor voltage is the PK value of the envelope signal.

(2) The detection of QP value is a special requirement in the conducted EMI measurement. The initial purpose of regulating the conducted EMI from CISPR is to guarantee the normal operation of telecommunication and broadcasting. The influence of the conducted EMI on telecommunication and broadcasting is mainly judged by the listening experiences of the audiences. An accidental, high-amplitude impulse may not affect the audience, while a repetitive signal with certain amplitude and pulsewidth may disrupt the telecommunication and broadcasting signals. Therefore, the QP detector is designed to gauge the intensity of a repetitive signal with certain magnitude and pulsewidth, and its output is closely correlated to the reaction of audience for audio signals. For the QP detector, the charging time constant τ_c is 1 ms and the discharging time constant τ_d is 160 ms. Obviously, τ_c is far shorter than τ_d [1]. When the charging and discharging is balanced, the QP value is acquired. Figure 2.8 shows the measured PK and QP values for the envelope signals with different repetitive period. Obviously, the PK value of these envelope signals is identical, and the QP value is related to the repetitive period of the envelope signals. The longer the repetitive period is, the lower the QP value will be. For the accidental signal, the resultant QP value is very small.

Fig. 2.8 The difference of detecting the PK and QP values

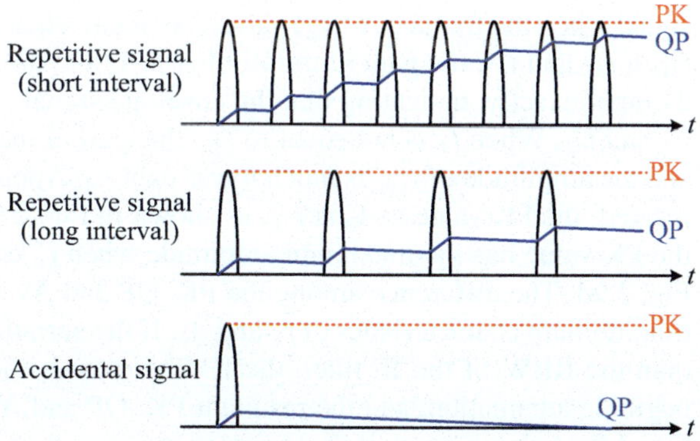

(3) The time constants for charging and discharing process of the AV detector are the same, and the AV detector outputs the average value of the envelope signal.

In EN 55022, the conducted EMI is only specified with the QP and AV values. However, the measurement of PK value is still essential, owing to the following reasons:

(1) Speed and Efficiency: The QP value is available until the charging and discharging process is balanced, so it takes long time to acquire the QP values. Comparatively, the PK detection is much faster, and the PK value is relatively close to the QP value. Therefore, the PK detection can be adopted to estimate the upper boundary of the QP value.
(2) Streamlined Testing: If the PK value is lower than the limit in the standard, the QP detection can be skipped for saving time because the QP value is lower than the PK value.
(3) Military Applications: In military applications, the standard normally specifies the PK value of the conducted EMI. This is due to the potential for a single pulse with sufficient amplitude to disrupt or even damage the sensitive instruments.

Figure 2.9 shows the PK, QP and AV values for various conducted EMI signals within the time interval $[t_0, t_1]$. It is important to note that the time interval, $[t_0, t_1]$, is chosen to be long enough to ensure that the charging and discharging processes of the detectors is balanced, enabling the accurate measurement of the PK, QP and AV values.

In Fig. 2.9a, b, the conducted EMI signal is sine wave with a fixed frequency of f_e. When f_e equals to the test frequency f_{IF} of the IF filter, the signal experiences a gain of 1, and it passes through the IF filter without attenuation. Therefore, the output of the envelope detector, v_{env}, tracks the amplitude A_e of its input exactly. In Fig. 2.9a, if the amplitude is fixed, the measured PK, QP and AV values are the same. In Fig. 2.9b, if the amplitude varies, the measured PK, QP and AV values satisfy the inequality PK > QP > AV. The difference between the PK, QP and AV values depends on the variation of the amplitude in the time-domain. The lower the amplitude variation is, the closer these detected values will be.

Figure 2.9c, d show the detected PK, QP and AV values of the conducted EMI signal when the frequency (f_e) varies. As seen, when f_e equals to the test frequency f_{IF} at t_{e1} and t_{e2}, the gain of the IF filter is 1, and the output of the IF filter follows its input exactly, indicating that the envelope signal v_{env} has the amplitude of A_e at t_{e1} and t_{e2}. When f_e is not equal to f_{IF}, the gain of the IF filter is far smaller than 1, and the amplitude of v_{env} is much lower than A_e. Apparently, when the amplitude A_e is fixed, the PK value is A_e at f_{IF}, as shown in Fig. 2.9c. If the amplitude A_e varies, the PK value has the maximum amplitude when f_e coincides with f_{IF}, as shown in Fig. 2.9d. The difference among the PK, QP and AV values are associated with the time-domain characteristics of f_e and A_e. If the variation range of f_e is much smaller than the RBW of the IF filter, the IF filter passes the conducted EMI signal with negilible attenuation, and the resultant PK, QP and AV values are similar to that in Fig. 2.9a, b. In contrast, if the variation range of f_e is much larger than the RBW of

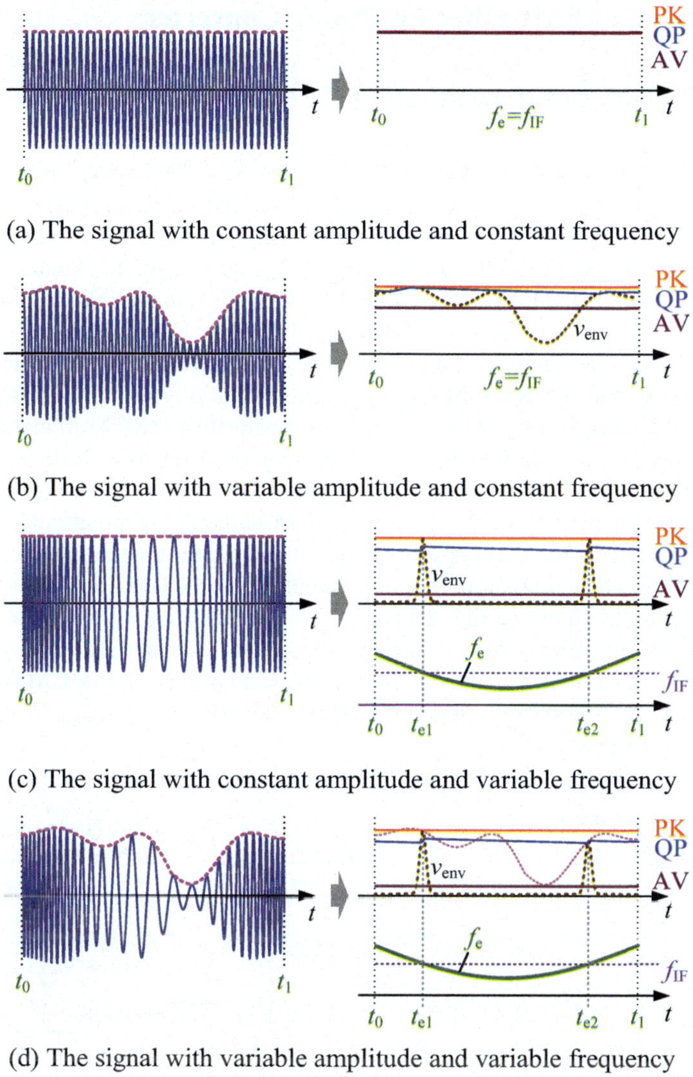

Fig. 2.9 The measured PK, QP and AV for various types of conducted EMI signals

the IF filter, v_{env} generally has low amplitude, and the AV value is much lower than the PK and QP values, as shown in Fig. 2.9c, d.

2.2 Design of EMI Filter for Power Converters

2.2.1 Topology and Attenuation Performance of EMI Filter

In order to comply with the conducted EMI standards, an EMI filter is usually inserted between the input power source and input side of the power converter. Taking the AC input as an example, Fig. 2.10 shows a typical EMI filter, where, L_{CM} is the CM inductor, L_{DM} is the DM inductor which utilizes the leakage inductance L_{lk} of the CM inductor, C_y is the CM capacitor, and C_{x1} and C_{x2} are the DM capacitors.

Referring to Fig. 2.10, since the CM noise propagates through both the input power cords with the same direction and amplitude, the DM capacitors have no effect on the CM noise, and the two CM capacitors are essentially in parallel. Besides, the leakage inductance of the CM inductor is far lower than the CM inductance. With these considerations, the CM filter can be extracted from Fig. 2.10, as shown in Fig. 2.11a. For the DM noise, the flux generated by the DM current in the CM inductors is cancelled by each other, so the CM inductor has no attenuation on the DM noise. Besides, the CM capacitance is much smaller than the DM capacitance due to the limit of the maximum leakage current. Thus, the DM filter can be extracted from Fig. 2.10, as shown in Fig. 2.11b.

The insert voltage gain (IVG) is usually adopted to describe the attenuation performance of the EMI filter, which is defined as the ratio of the original CM (DM) noise v_{CM_ori} (v_{DM_ori}) and the attenuated CM (DM) noise v_{CM_att} (v_{DM_att}) with the CM (DM) filter, expressed as

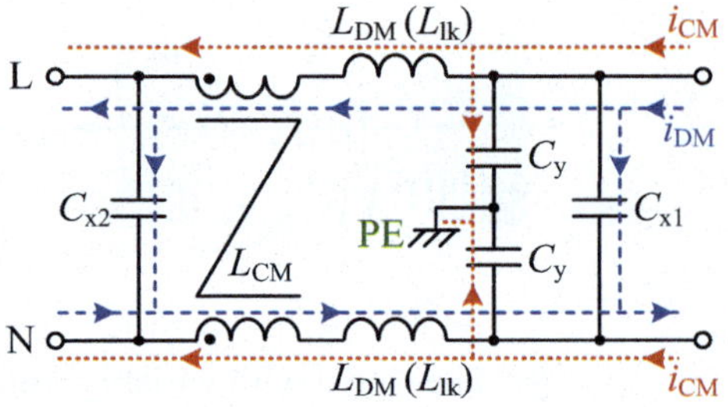

Fig. 2.10 A typical EMI filter

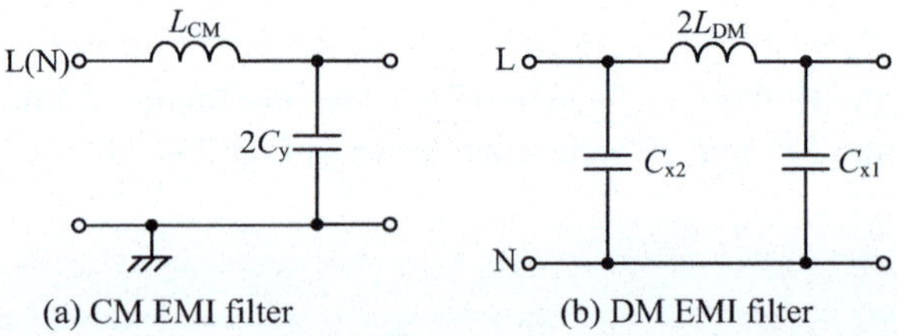

Fig. 2.11 The CM and DM EMI filters extracted from the EMI filter in Fig. 2.10

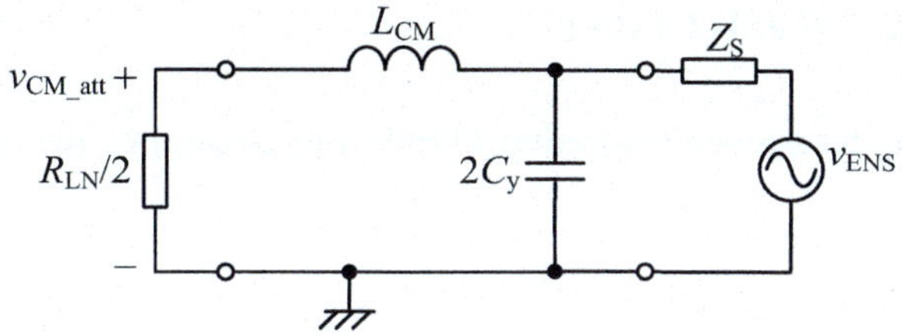

Fig. 2.12 The EMI filter with the CM noise equivalent circuit of the converter and the CM impedance from the LISN side

$$|\text{IVG}|_{\text{CM}} = \left|\frac{v_{\text{CM\_ori}}}{v_{\text{CM\_att}}}\right|, \quad |\text{IVG}|_{\text{DM}} = \left|\frac{v_{\text{DM\_ori}}}{v_{\text{DM\_att}}}\right| \tag{2.8}$$

Based on Fig. 2.11a, the CM filter with the CM impedance at the LISN side and the CM noise equivalent circuit of the power converter is provided in Fig. 2.12. The CM impedance at the LISN side is $R_{LN}/2$ since the test resistors of the L line and N line are essentially in parallel when measuring the CM noise. For the power converter, the CM noise equivalent circuit can be simplified as a Thevenin's equivalent circuit [4], which is represented by an equivalent noise source (ENS) v_{ENS} and a CM impedance Z_S connected in series.

According to Fig. 2.12, the original CM noise v_{CM_ori} without the EMI filter is given by

$$v_{\text{CM\_ori}} = \frac{R_{\text{LN}}/2}{Z_{\text{S}} + R_{\text{LN}}/2} v_{\text{ENS}} = \frac{R_{\text{LN}}}{2Z_{\text{S}} + R_{\text{LN}}} v_{\text{ENS}} \tag{2.9}$$

With the EMI filter, the attenuated CM noise v_{CM_att} is calculated as

$$\begin{aligned} v_{\text{CM\_att}} &= \frac{R_{\text{LN}}/2}{sL_{\text{CM}} + R_{\text{LN}}/2} \frac{(sL_{\text{CM}} + R_{\text{LN}}/2)//\frac{1}{2sC_y}}{Z_{\text{S}} + (sL_{\text{CM}} + R_{\text{LN}}/2)//\frac{1}{2sC_y}} v_{\text{ENS}} \\ &= \frac{R_{\text{LN}}}{4L_{\text{CM}}C_y Z_{\text{S}} s^2 + 2\left(L_{\text{CM}} + C_y Z_{\text{S}} R_{\text{LN}}\right)s + 2Z_{\text{S}} + R_{\text{LN}}} v_{\text{ENS}} \end{aligned} \tag{2.10}$$

By combining (2.8)–(2.10), the IVG of the CM filter is derived as

$$|\text{IVG}|_{\text{CM}} = \left|\frac{4L_{\text{CM}}C_y Z_{\text{S}} s^2 + 2\left(L_{\text{CM}} + C_y Z_{\text{S}} R_{\text{LN}}\right)s + 2Z_{\text{S}} + R_{\text{LN}}}{2Z_{\text{S}} + R_{\text{LN}}}\right| \tag{2.11}$$

The IVG of the DM filter can be derived in a similar way, and it is not given here.

2.2.2 Design of EMI Filter

Figure 2.13 outlines a general procedure of EMI filter design [4]. To begin, the original CM or DM noises of the power converter are determined, either through prediction or measurement. The attenuation requirements for the CM and DM filters are then established by substracting the conducted EMI limits from the CM or DM noises, and an appropriate margin is also considered. It should be noted that, the conduced EMI spectra of the power converter can be quite different at various working conditions. Therefore, the attenuation requirement should be calculated based on the worst-case conducted EMI spectrum.

The worst-case conducted EMI spectrum is characterized by the lowest corner frequency of the EMI filter, which corresponds to the largest volume/weight of the EMI filter [5]. As shown in Fig. 2.14, the corner frequency of the EMI filter for Case II is lower than that of the EMI filter for Case I. Consequently, the conducted EMI spectrum for Case II is worse.

Secondly, the EMI filter topology can be selected by considering the CM and DM impedances of the converter and the LISN side [6]. According to the attenuation requirement, the parameters of the filter can be calculated.

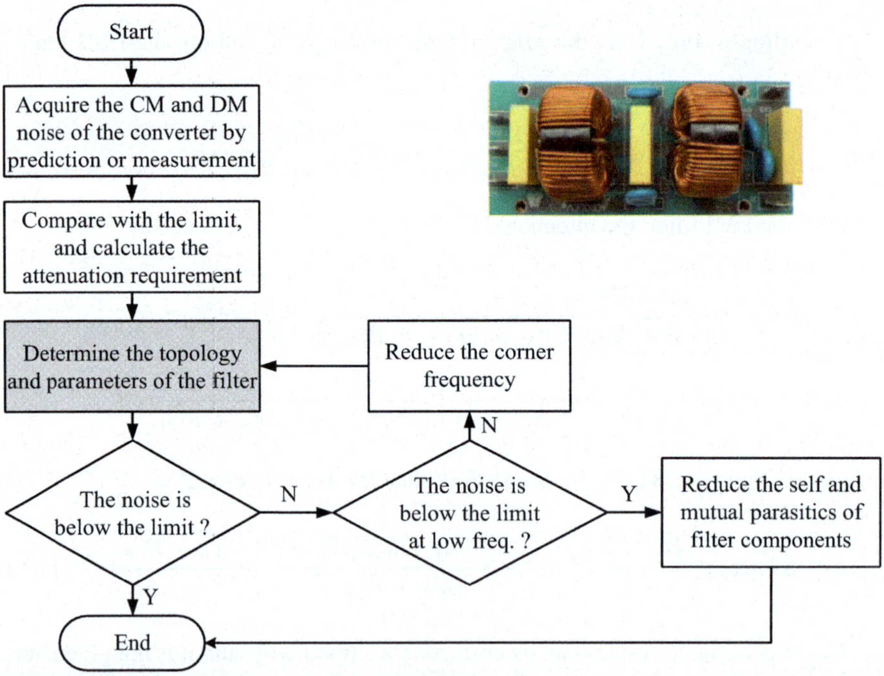

Fig. 2.13 The basic procedure for the EMI filter design

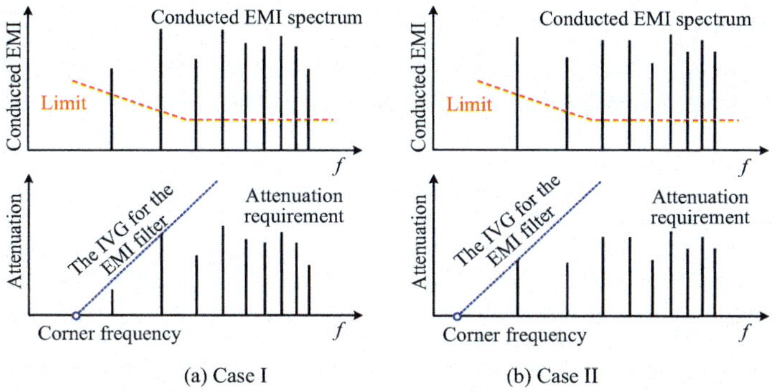

Fig. 2.14 The conducted EMI spectrum and its attenuation requirement for different case

Finally, the attenuated conducted EMI of the converter can be measured by inserting the designed CM and DM EMI filters. In cases where the conducted EMI levels are still higher than the limits at low frequencies, the attenuation can be enhanced by increasing the filter inductance and capacitance. However, as the inductance or capacitance goes larger, the parasitics will be amplified. For example, multi-turn and multi-layer inductors often exhibit substantial parasitic capacitance [7–9], which can compromise filter performance at higher frequencies, making the noise level unexpectedly higher. In order to improve the attenuation performance of EMI filter at higher frequencies, efforts should be put into the minimization or elimination of the self and mutual parasitics of the components. This can be achieved through strategic winding arrangements for inductors, adjustment in the relative position of components, or the incorporation of impedance cancellation techniques [10–18].

2.3 Summary

In this chapter, the main instruments for measuring the conducted EMI are introduced. The LISN is used to provide specific impedance within the conducted EMI frequency range, ensuring repeatability in measuring the conducted EMI. The EMI receiver picks up the EMI signal from the LISN, and outputs the corresponding PK, QP, and AV spectra. Additionally, the noise separator can be adopted to separate the conducted EMI into CM and DM noises, which greatly facilitates the diagnosis of the conducted EMI spectrum and aids in the design of EMI filters.

This chapter also presents the design of EMI filter. The attenuation requirement is calculated by substracting the conducted EMI limits from the original conducted EMI spectrum of power converters. The parameters of the components are then calculated according to the attenuation requirement and the IVG of the EMI filter.

References

1. CISPR 16-1-2, *Specification for Radio Disturbance and Immunity Measuring Apparatus and Methods—Part 1–2* (British Standards, British, 2009)
2. W. Schaefer, Significance of EMI receiver specifications for commercial EMI compliance testing, in *Proceedings of International Symposium on Electromagnetic Compatibility (APEMC)* (2004), pp. 741–746
3. S. Wang, F.C. Lee, W.G. Odendaal, Characterization, evaluation, and design of noise separator for conducted EMI noise diagnosis. IEEE Trans. Power Electron. **20**(4), 974–982 (2005)
4. F.Y. Shih, D. Chen, Y. Wu, Y. Chen, A procedure for designing EMI filters for ac line applications. IEEE Trans. Power Electron. **11**(1), 170–181 (1996)
5. K. Mainali, R. Oruganti, K. Viswanathan, A metric for evaluating the EMI spectra of power converters. IEEE Trans. Power Electron. **23**(4), 2075–2081 (2008)
6. S. Wang, Characterization and cancellation of high-frequency parasitics for EMI filters and noise separators in power electronics applications. PhD Dissertation, Virginia Polytechnic Institute and State University, Blacksburg, USA, 2005
7. A. Nagel, R.W. De Doncker, Systematic design of EMI-filters for power converters, in *Proceedings of Annual Conference of IEEE Industry Applications Society (IAS)* (2000), pp. 2523–2525
8. M. Kumar, V. Agarwal, Power line filter design for conducted electromagnetic interference using time-domain measurements. IEEE Trans. Electromagn. Compat. **48**(1), 178–186 (2006)
9. Y.-S. Lai, P.S. Chen, New EMI filter design method for single phase power converter using software-based noise separation method, in *Proceedings of Annual Conference of IEEE Industry Applications Society (IAS)* (2007), pp. 2282–2288
10. S. Wang, F.C. Lee, D. Chen, Effects of parasitic parameters on EMI filter performance. IEEE Trans. Power Electron. **19**(3), 869–877 (2004)
11. S. Wang, R. Chen, J.D. Van Wyk, Developing parasitic cancellation technologies to improve EMI filter performance for switching mode power supplies. IEEE Trans. Electromagn. Compat. **48**(4), 921–929 (2005)
12. S. Wang, F.C. Lee, W.G. Odendaal, Improvement of EMI filter performance with parasitic coupling cancellation. IEEE Trans. Power Electron. **20**(5), 1221–1228 (2005)
13. S. Wang, F.C. Lee, W.G. Odendaal, Characterization and parasitic extraction of EMI filters using scattering parameters. IEEE Trans. Power Electron. **20**(2), 502–510 (2005)
14. S. Wang, F.C. Lee, W.G. Odendaal, Cancellation of capacitor parasitic parameters for noise reduction application. IEEE Trans. Power Electron. **21**(4), 1125–1132 (2006)
15. S. Wang, F.C. Lee, J.D. Van Wyk, Design of inductor winding capacitance cancellation for EMI suppression. IEEE Trans. Power Electron. **21**(6), 1825–1832 (2006)
16. S. Wang, F.C. Lee, J.D. Van Wyk, Inductor winding capacitance cancellation using mutual capacitance concept for noise reduction application. IEEE Trans. Electromagn. Compat. **48**(2), 311–318 (2006)
17. S. Wang, F.C. Lee, J.D. Van Wyk, A study of integration of parasitic cancellation techniques for EMI filter design with discrete components. IEEE Trans. Power Electron. **23**(6), 3094–3102 (2008)
18. S. Wang, F.C. Lee, Analysis and applications of parasitic capacitance cancellation techniques for EMI suppression. IEEE Trans. Industr. Electron. **57**(9), 3109–3117 (2010)

Chapter 3
Common-Mode and Differential-Mode Noise Equivalent Circuits for Boost Power Factor Correction Converter

3.1 Common-Mode and Differential-Mode Noises in Boost PFC Converter

3.1.1 Noise Propagation Path and Mixed-Mode Noise Reduction

In this section, the propatation paths of the boost power factor correction (PFC) converter are first investigated, in order to analyze its conducted electromagnetic interference (EMI). Figure 3.1 shows the measurement of conducted EMI for boost PFC converters, where, $D_1 \sim D_4$ constitute the rectifier bridge, and boost inductor L_b, power switch Q_b, diode D_b and output filter capacitor C_o constitute the boost converter. C_p is the parasitic capacitance between the drain of Q_b and the protective earth (PE). The line impedance stabilization network (LISN) is placed between the AC input voltage source and the boost PFC converter, which provides specific impedance for the conducted EMI within [150 kHz, 30 MHz]. For the convenience of illustration, the LISN is simplified as a 50-Ω resistor R_{LN} in the following.

The conducted EMI propagation path of boost PFC converter is illustrated in Fig. 3.2, where, the drain-to-source voltage of power switch Q_b, referred to as v_{DS}, is treated as the noise voltage source. i_{Lb} is boost inductor current, and i_{Cp} is the current flowing through C_p. As seen in Fig. 3.2a, during the positive half line cycle, both D_1 and D_4 are conducting, i_{Lb} flows through the two test resistors R_{LN} in the LISN, while i_{Cp} only flows through R_{LN} in the N line and returns to v_{DS} through D_4.

© The Author(s), under exclusive license to Springer Nature Singapore Pte Ltd. 2024

X. Ruan et al., *Conducted Electromagnetic Interference in Power Converters: Modeling, Prediction and Reduction*, CPSS Power Electronics Series,
https://doi.org/10.1007/978-981-99-9295-9_3

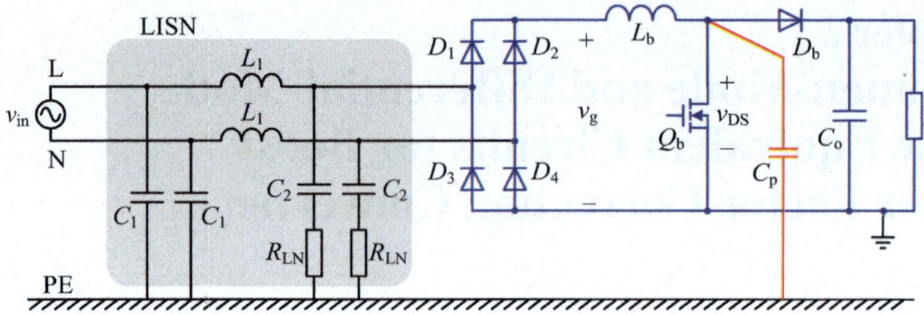

Fig. 3.1 Conducted EMI measurement for boost PFC converter

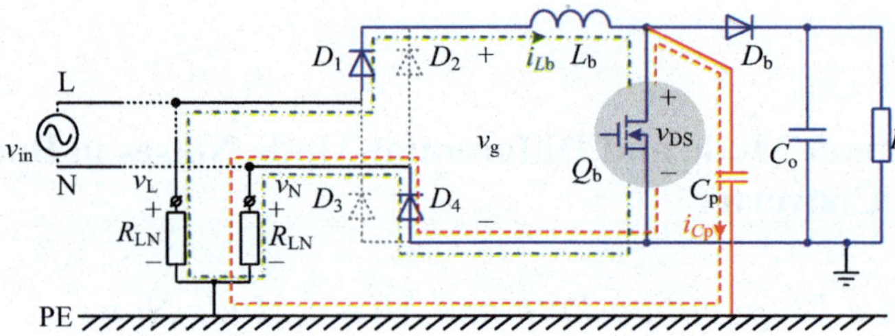

(a) D_1 and D_4 are conducting (positive half line cycle)

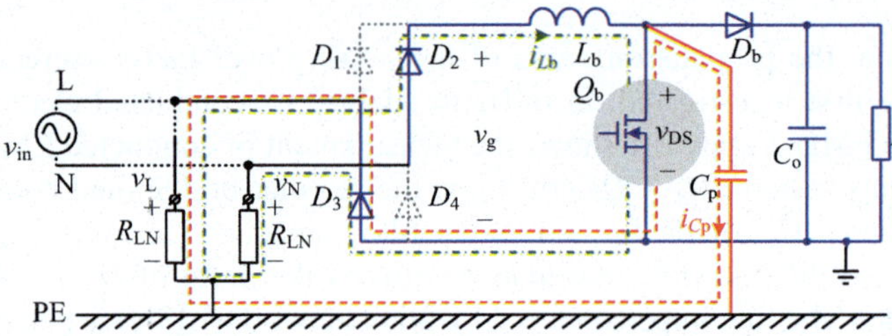

(b) D_2 and D_3 are conducting (negative half line cycle)

Fig. 3.2 The conducted EMI propagation path of boost PFC converter

According to Fig. 3.2a, the noise voltages across the test resistors of the L line and N lines are expressed as

$$\begin{cases} v_{\mathrm{L}} = -50 i_{Lb} \\ v_{\mathrm{N}} = 50\left(i_{Lb} - i_{Cp}\right) \end{cases} \quad (v_{\mathrm{in}} \geq 0) \tag{3.1}$$

According to (3.1), the common-mode (CM) and differential mode (DM) voltages are

$$\begin{cases} v_{\mathrm{CM}} = 0.5(v_{\mathrm{L}} + v_{\mathrm{N}}) = 25 i_{\mathrm{p}} \\ v_{\mathrm{DM}} = 0.5(v_{\mathrm{L}} - v_{\mathrm{N}}) = 25\left(2 i_{Lb} - i_{Cp}\right) \end{cases} \quad (v_{\mathrm{in}} \geq 0) \tag{3.2}$$

As shown in Fig. 3.2b, during the negative half line cycle, both D_2 and D_3 are conducting, while i_{Cp} only flows through R_{LN} in the L line and returns to v_{DS} through D_3. The noise voltages in the L line and N lines are expressed as

$$\begin{cases} v_L = -50(i_{Lb} + i_{Cp}) \\ v_N = 50 i_{Lb} \end{cases} \quad (v_{in} < 0) \tag{3.3}$$

According to (3.3), the corresponding CM and DM voltages can be derived as

$$\begin{cases} v_{CM} = 0.5(v_L + v_N) = 25 i_{Cp} \\ v_{DM} = 0.5(v_L - v_N) = 25(2 i_{Lb} + i_{Cp}) \end{cases} \quad (v_{in} < 0) \tag{3.4}$$

Referring to (3.2) and (3.4), it is observed that v_{CM} is only related to i_{Cp}, so i_{Cp} is regarded as the CM noise current. However, v_{DM} is related to i_{Lb} and i_{Cp}, and it changes with the polarity of the input voltage. Therefore, the DM noise is partly contributed by the CM noise current, leading to the mixed-mode noise. Besides, when the inductor current becomes discontinuous, the resonance between boost inductor L_b and the parasitic capacitance of Q_b and $D_1 \sim D_4$ will occur, causing voltage spike in the input rectified voltage v_g. Such voltage spike will not only damage the rectifier diode, but also introduce extra CM noise [1].

1. Introducting the DM Capacitor C_{xdc}

To eliminate the voltage spikes in v_g, a DM capacitor C_{xdc} can be added at the DC side of the rectifier bridge, as shown in Fig. 3.3. C_{xdc} is usually much larger than the capacitance of Q_b and $D_1 \sim D_4$, so the voltage of C_{xdc} is almost unchanged when the inductor current becomes discontinuous. When the rectifier bridge is in the on-state, C_{xdc} offers a low impedance path between the L line and N line. Thus, i_{Cp} is able to flow through R_{LN} in both lines, avoiding the conversion of the CM noise into the DM noise. Therefore, C_{xdc} also reduces the mixed-mode noise during the on-state of the rectifier bridge.

During the zero-crossing of the AC input voltage, the input power of the converter is mainly supplied by C_{xdc}, as shown in Fig. 3.4 with the shaded area, and the rectifier bridge is turned off. This will lead to distortion in the input current [2, 3]. Therefore, C_{xdc} cannot be too large, and it is selected to make its impedance be far lower than 50 Ω at the frequencies above 150 kHz. For instance, a 0.2-μF C_{xdc} is enough since its impedance is 5 Ω at 150 kHz.

2. Introducing the DM Capacitor C_{xac}

Figure 3.5 shows the CM current path during the off-state of the rectifier bridge. During the positive half line cycle, when C_p is charged, i_{Cp} flows through D_1 and generates the noise voltage across R_{LN} in the L line; when C_p is discharged, i_{Cp} flows through R_{LN} in the N line and D_4. During the negative half line cycle, when C_p is charged, i_{Cp} flows through R_{LN} in the N line and D_2; when C_p is discharged, i_{Cp} flows through R_{LN} in the L line and D_3.

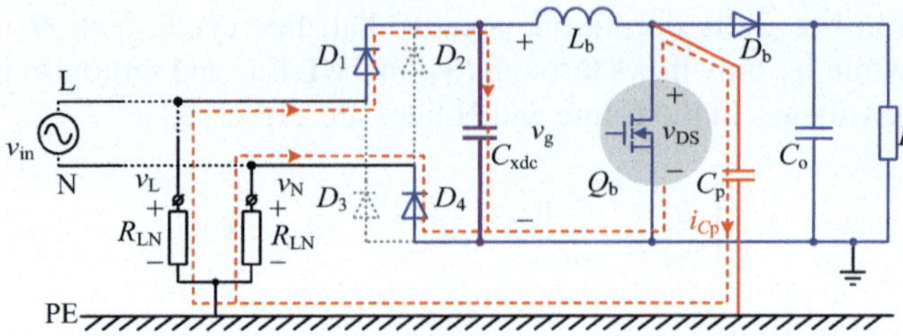

(a) D_1 and D_4 are conducting (positive half line cycle)

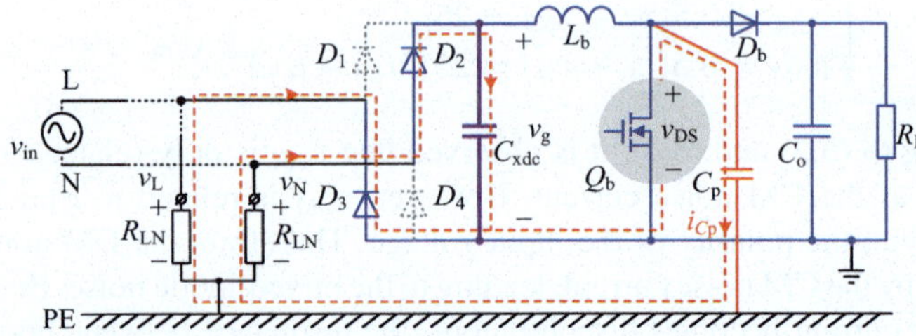

(b) D_2 and D_3 are conducting (negative half line cycle)

Fig. 3.3 The influence of C_{xdc} on the CM current path

Fig. 3.4 The off-state of rectifier bridge with C_{xdc}

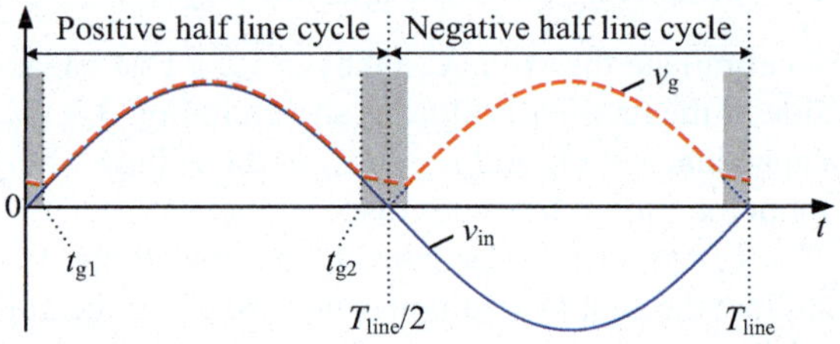

As seen in Fig. 3.5, since i_{Cp} does not flow through the R_{LN} in both lines, the DM noise is created, which is the mixed-mode noise during the off-state of the rectifier bridge [4–6]. Although the duration of off-state is quite short, the mixed-mode noise appears during every zero-crossing of the input voltage, and it is a signal with long repetitive period that generates high peak (PK) and quasi-peak (QP) values. To address this issue, a DM capacitor C_{xac} is added at the AC side of the rectifier bridge, providing a low impedance path for both lines, as shown in Fig. 3.6. In this way, i_{Cp} can flow through the R_{LN} in both lines.

In summary, by adding the DM capacitors to both the AC and DC sides of the rectifier bridge, the mixed-mode noise of boost PFC converters can be suppressed.

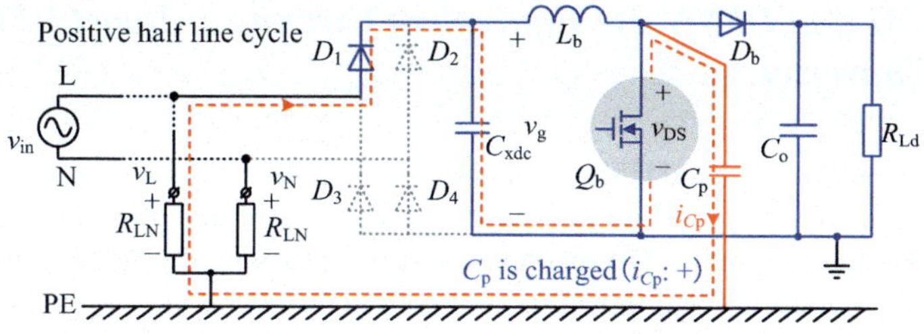

(a) C_p is charged (i_{Cp} is positive) with D_1 turned on

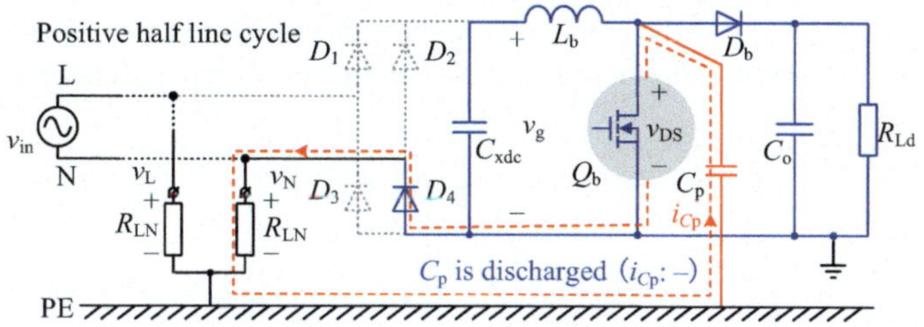

(b) C_p is discharged (i_{Cp} is negative) with D_1 turned on

Fig. 3.5 The CM current path during the off-state of rectifier bridge with C_{xdc}

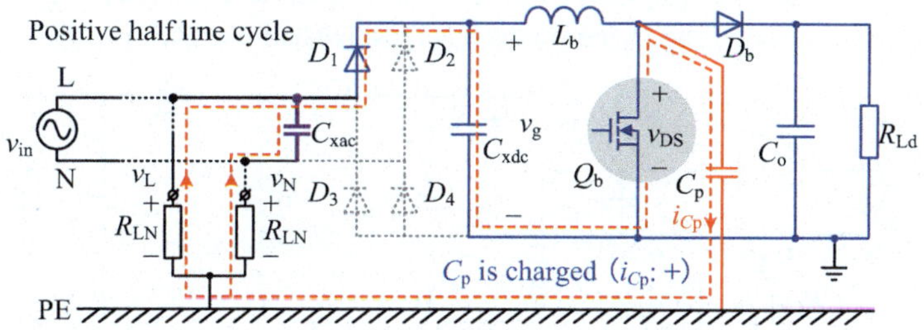

(a) C_p is charged (i_{Cp} is positive)

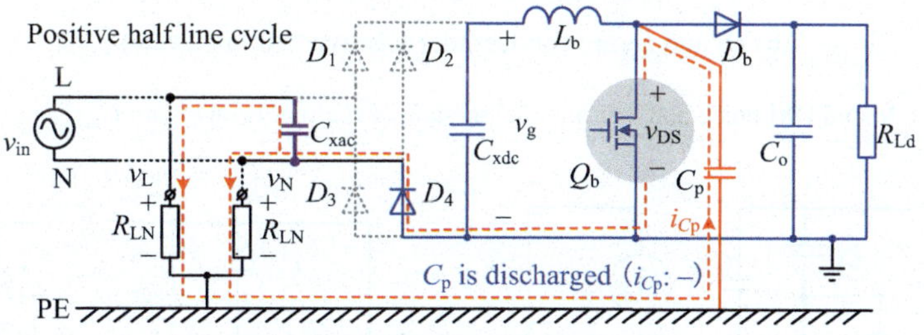

(b) C_p is discharged (i_{Cp} is negative)

Fig. 3.6 The CM current path during the off-state of rectifier bridge with C_{xdc} and C_{xac}

3.1.2 CM and DM Noise Equivalent Circuits of Boost PFC Converter

Figure 3.7 gives the CM and DM noise propagation paths of boost PFC converter with the DM capacitors C_{xdc} and C_{xac}. By extracting the associated branches in the CM and DM propagation paths, the resultant circuits are shown in Fig. 3.8, where, C_{xdc} and C_{xac} are in parallel, and they are combined and denoted as C_{x1} in the following discussion.

At the frequencies higher than 150 kHz, the impedance of C_{x1} is far lower than R_{LN} (50 Ω), and C_{x1} can be treated as a short circuit in the CM noise equivalent circuit. Thus, the circuit shown in Fig. 3.8a can be simplified, as shown in Fig. 3.9a.

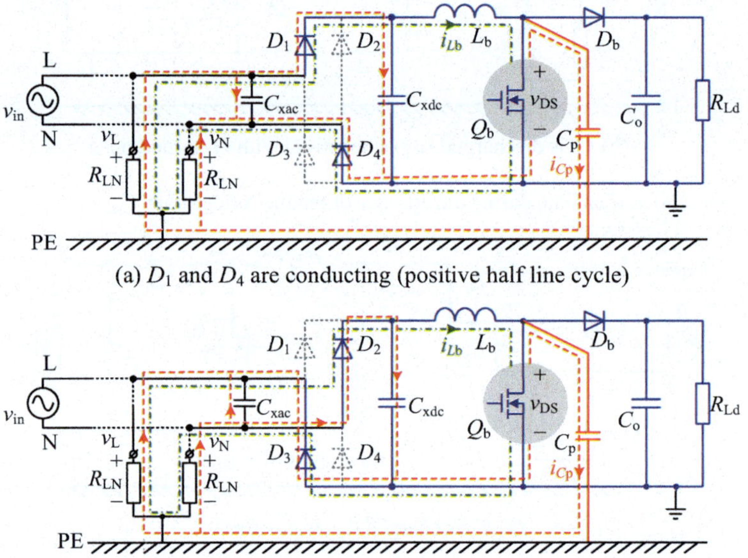

(a) D_1 and D_4 are conducting (positive half line cycle)

(b) D_2 and D_3 are conducting (negative half line cycle)

Fig. 3.7 CM and DM noise paths during the on-state of rectifier bridge with C_{xdc} and C_{xac}

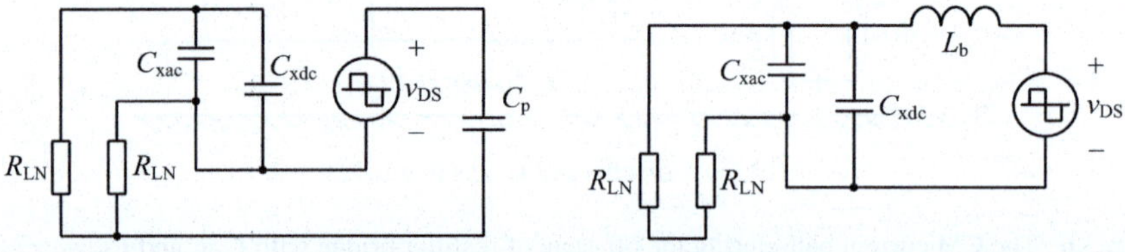

(a) CM noise related branches (b) DM noise related branches

Fig. 3.8 The CM and DM noise related branches in boost PFC converter

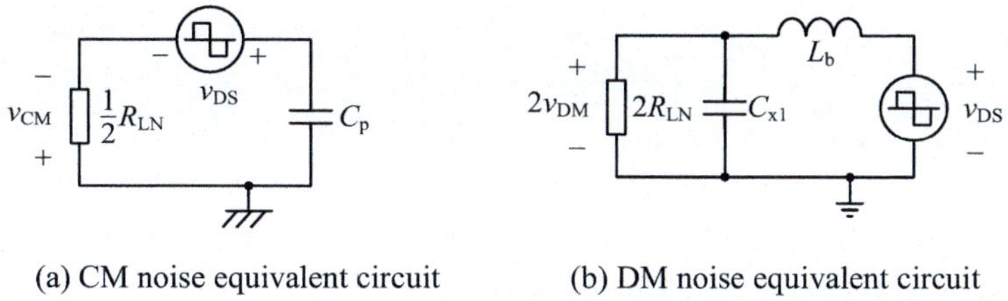

Fig. 3.9 CM and DM noise equivalent circuit of boost PFC converter

Likewise, the circuit shown in Fig. 3.8b can also be simplified, as shown in Fig. 3.9b. Note that the voltage across $R_{LN}/2$ is v_{CM}, and the voltage across $2R_{LN}$ is $2v_{DM}$.

Denoting the harmonic amplitude of v_{DS} as V_{DS_k} (k is the harmonic order), the harmonic amplitude of the CM and DM noises, referred to as V_{CM_k} and V_{DM_k}, are expressed as

$$\begin{cases} V_{CM\_k} = |\mathrm{CMTG}(kf_s)| \cdot V_{DS\_k} \\ V_{DM\_k} = |\mathrm{DMTG}(kf_s)| \cdot V_{DS\_k} \end{cases} \tag{3.5}$$

where, |CMTG(f)| and |DMTG(f)| are the CM and DM voltage transfer gain, respectively. According to Fig. 3.9, |CMTG(f)| and |DMTG(f)| are given by

$$|\mathrm{CMTG}(f)| = \frac{\pi f C_p R_{LN}}{\sqrt{(\pi f C_p R_{LN})^2 + 1}} \tag{3.6}$$

$$|\mathrm{DMTG}(f)| = \frac{R_{LN}}{2\sqrt{R_{LN}^2(4\pi^2 f^2 L_b C_{x1} - 1)^2 + (4\pi f L_b)^2}} \tag{3.7}$$

Since C_p is small, $\pi f C_p R_{LN}$ is far smaller than 1, and (3.6) can be approximated as

$$|\mathrm{CMTG}(f)| = \pi f C_p R_{LN} \tag{3.8}$$

Obviously, |CMTG(f)| increases with the increase of frequency, and the slope is 20 dB/dec.

For the DM filter, at the frequencies higher than 150 kHz, $4\pi^2 f^2 L_b C_{x1}$ is usually much larger than 1, and $4\pi^2 f^2 L_b C_{x1} R_{LN}$ is far larger than $\pi f L_b$. Therefore, (3.7) can be approximated as

$$|\mathrm{DMTG}(f)| = \frac{1}{8\pi^2 f^2 L_b C_{x1}} \tag{3.9}$$

Based on (3.9), |DMTG(f)| decreases with the increase of frequency, and the slope is − 40 dB/dec.

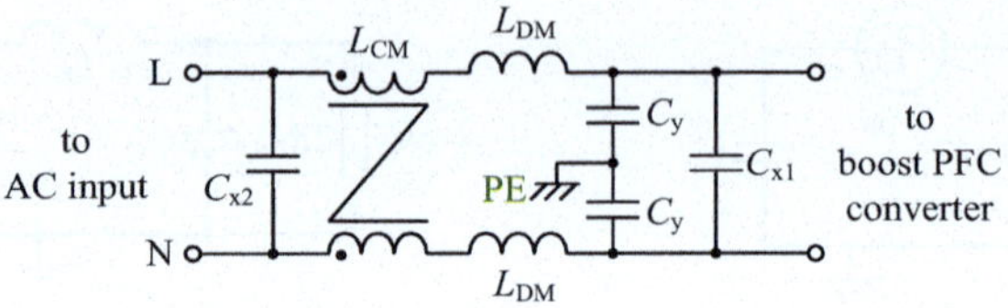

Fig. 3.10 The EMI filter for boost PFC converter

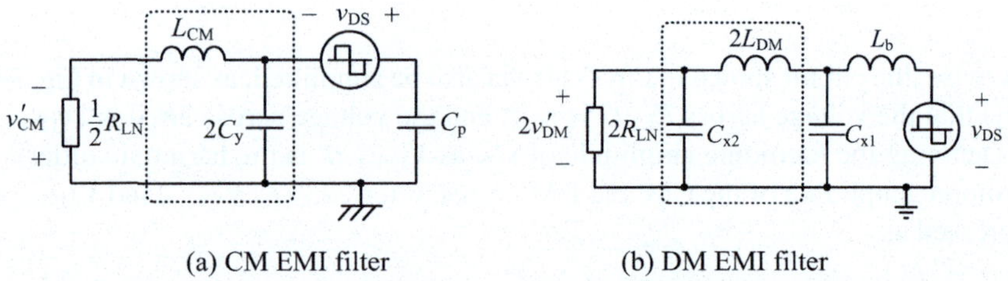

Fig. 3.11 The boost PFC converter with the CM and DM EMI filters

3.2 Design of the EMI Filter for Boost PFC Converters

3.2.1 EMI Filter for Boost PFC Converters

The EMI filter suitable for boost PFC converters is shown in Fig. 3.10 [7], where, the CM filter is realized by the CM inductor L_{CM} and the CM capacitor C_y, while the DM filter is consisted of the DM inductor L_{DM} and the DM capacitors C_{x1} and C_{x2}. The CM and DM filters extracted from Fig. 3.10 are given in Fig. 2.11.

With the CM and DM filters in Fig. 2.11, as well as the CM and DM noise equivalent circuits in Fig. 3.9, Fig. 3.11a, b show the boost PFC converter with the CM and DM filters, respectively.

3.2.2 Design of the EMI Filter for the Boost PFC Converter

According to the attenuation requirement and the insert voltage gain (IVG) curve for the CM and DM noises, the corner frequency f_c of the IVG curve and the tagency point of these two curves can be found, as shown in Fig. 3.12. According to Fig. 3.11, both the CM and DM EMI filters are second order ones, so their IVG curves above f_c are both 40 dB/dec.

In Fig. 3.12, the frequency at the tagency point is denoted as f_{req}, and the attenuation requirement of the tagency point is referred to as A_{req}. According to the highlighted triangle, we have

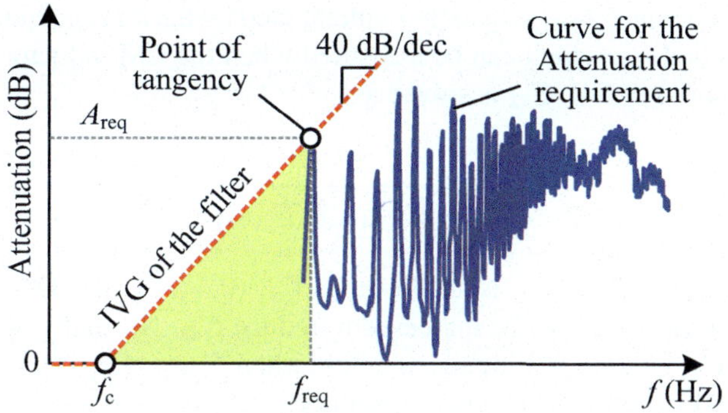

Fig. 3.12 The attenuation requirement and the IVG of EMI filter for conducted EMI

$$A_{\mathrm{req}} = 40\left(\lg f_{\mathrm{req}} - \lg f_{\mathrm{c}}\right) \tag{3.10}$$

According to (3.10), f_{c} is solved as

$$f_{\mathrm{c}} = f_{\mathrm{req}} \cdot 10^{-\frac{A_{\mathrm{req}}}{40}} \tag{3.11}$$

To determine the component parameters for the CM filter shown in Fig. 3.11a, the CM capacitance is first calculated by considering the leakage current limit [8]. The corner frequency $f_{\mathrm{c\_CM}}$ is expressed as

$$f_{\mathrm{c\_CM}} = \frac{1}{2\pi\sqrt{L_{\mathrm{CM}}\left(2C_{\mathrm{y}}\right)}} \tag{3.12}$$

Thus, the CM inductance can be derived as

$$L_{\mathrm{CM}} = \frac{1}{2C_{\mathrm{y}}\left(2\pi\, f_{\mathrm{c\_CM}}\right)^2} \tag{3.13}$$

When designing the DM filter for boost PFC converters, the DM capacitors C_{xdc} and C_{xac} are first added at the DC and AC sides of the rectifier bridge for suppressing the mixed-mode noise. As discussed earlier, C_{xdc} and C_{xac} should be larger than 0.2 μF, and C_{x1}, which equals to $C_{\mathrm{xdc}} + C_{\mathrm{xac}}$, can be preselected as 0.4 μF. Referring to Fig. 3.11b, the corner frequency $f_{\mathrm{c\_DM}}$ of the DM filter is calculated as

$$f_{\mathrm{c\_DM}} = \frac{1}{2\pi\sqrt{2L_{\mathrm{DM}}C_{\mathrm{x2}}}} \tag{3.14}$$

The DM inductance L_{DM} is usually implemented by the leakage inductance L_{lk} of the CM inductor L_{CM}, which can be measured when the CM inductor is fabricated. Thus, according to (3.14), C_{x2} is solved as

$$C_{x2} = \frac{1}{2L_{DM}(2\pi f_{c\_DM})^2} \tag{3.15}$$

According to (3.15), if C_{x2} is smaller than C_{x1} (0.4 μF), the DM filter design is finished. Otherwise, C_{x1} can be increased to reduce C_{x2}. Normally, the overall DM capacitance ($C_{x1} + C_{x2}$) will be the smallest when $C_{x1} = C_{x2}$ [9], resulting in the smallest overall volume for the DM capacitors.

3.3 Summary

In this chapter, the propagation paths for the CM and DM noises of boost PFC converters are analyzed. It is pointed out that the CM and DM noises affect each other, leading to the mixed-mode noise. To suppress the mixed-mode noise, the DM capacitors are added at the DC and AC sides of the rectifier bridge, eliminating the mixed-mode noise during the on-state and off-state of the rectifier bridge, respectively. After that, the CM and DM noise equivalent circuits of boost PFC converters are derived, and the design of EMI filters is given. With the derived CM and DM noise equivalent circuits in this chapter, the conducted EMI spectra for boost PFC converters will be explored in Chaps. 4 and 5, specially investigating their behavior under different control schemes.

References

1. P. Kong, Y. Jiang, F.C. Lee, Common mode EMI noise characteristics of low-power ac-dc converters. IEEE Trans. Power Electron. **27**(2), 731–738 (2012)
2. J. Sun, On the zero-crossing distortion in single-phase PFC converters. IEEE Trans. Power Electron. **9**(3), 685–692 (2004)
3. W. Levron, H. Kim, R.W. Erickson, Design of EMI filters having low harmonic distortion in high-power-factor converters. IEEE Trans. Power Electron. **29**(7), 3403–3413 (2014)
4. J. Meng, W. Ma, A new technique for modeling and analysis of mixed-mode conducted EMI noise, in *Proceedings of IEEE Power Electronics Specialists Conference (PESC)* (2004), pp. 1034–1042
5. H.-I. Hsieh, D.-Y. Chen, EMI filter design method incorporating mix-mode conducted noise for off-line applications, in *Proceedings of Vehicle Power and Propulsion Conference (VPPC)* (2008), pp. 1617–1622
6. H.-I. Hsieh, Effects of mix-mode noise emissions on the design method of power factor correction boost rectifier EMI filters, in *Proceedings of International Power Electronics Conference (IPEC)* (2010), pp. 2438–2443

7. S. Wang, Characterization and cancellation of high-frequency parasitics for EMI filters and noise separators in power electronics applications. Ph.D. Dissertation, Virginia Polytechnic Institute and State University, Blacksburg, USA, 2005
8. IEC 60990, *Methods of Measurement of Touch Current and Protective Conductor Current* (British Standard, British, 1999)
9. Q. Ji, Research on conducted EMI of boost PFC converter. Ph.D Dissertation, Nanjing University of Aeronautics and Astronautics, Nanjing, 2014

Chapter 4
Prediction of the Conducted Electro-magnetic Interference for Average-Current-Controlled Boost Power Factor Correction Converter

4.1 Operation Modes of Average-Current-Controlled Boost PFC Converters

Figure 4.1 shows the boost power factor correction (PFC) converter with average-current-control, where, $D_1 \sim D_4$ constitute the input bridge rectifier; boost inductor L_b, power switch Q_b, diode D_b, and output capacitor C_o constitute the boost converter; and R_{Ld} is the load resistor, and C_{xac} and C_{xdc} are the capacitors for suppressing the mixed-mode noise.

The control scheme includes the output voltage closed-loop and the inductor current closed-loop. The output voltage V_o is sensed and compared with the reference signal V_{o_ref}, and the error signal is sent into the voltage regulator $G_v(s)$. The output of the voltage regulator is multiplied with the sensed rectified input voltage, which serves as the current reference signal i_{Lb_ref}. The inductor current i_{Lb} is sensed and compared with i_{Lb_ref}, and the error signal is sent to the current regulator $G_i(s)$. The output of $G_i(s)$ is compared with the sawtooth carrier, generating the driving signals for power switch Q_b.

Under different input voltage and load conditions, the boost PFC converter will be operated either in continuous conduction mode (CCM) or discontinuous conduction mode (DCM) in a switching cycle. Thus, the boost PFC converter may operate in all-CCM, part-CCM/DCM and all-DCM [1–4]. Figure 4.2 shows the waveforms of boost inductor current in different operation modes, where, T_{line} is the line period.

The rectified input voltage of boost PFC converter can be expressed as

$$v_g(t) = |v_{in}(t)| = \sqrt{2}V_{in}|\sin \omega_{in}t| \tag{4.1}$$

where, v_{in} is the input voltage, V_{in} is the root-mean-square (rms) value of the input voltage, and ω_{in} is the angular frequency of the input voltage.

© The Author(s), under exclusive license to Springer Nature Singapore Pte Ltd. 2024
X. Ruan et al., *Conducted Electromagnetic Interference in Power Converters: Modeling, Prediction and Reduction*, CPSS Power Electronics Series,
https://doi.org/10.1007/978-981-99-9295-9_4

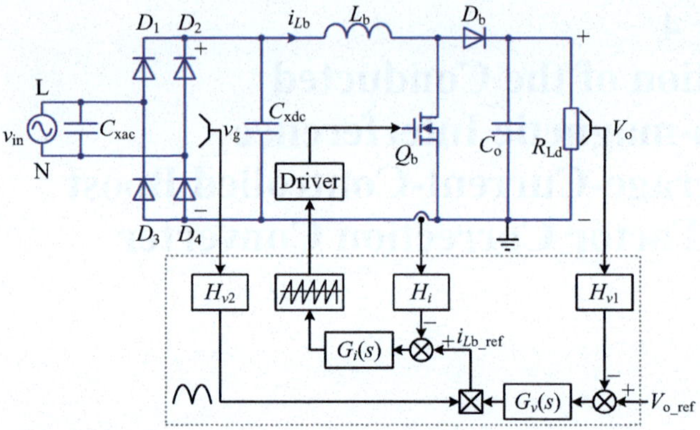

Fig. 4.1 Average-current-controlled boost PFC converter

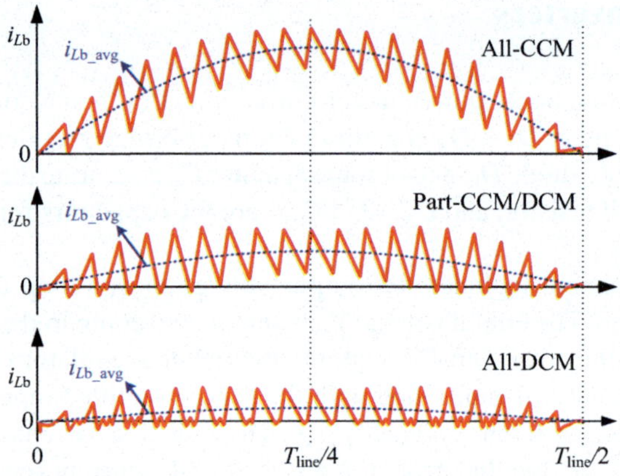

Fig. 4.2 Boost inductor current waveforms under different operation modes

When operated in CCM, the duty cycle D_{y_CCM} and the inductor current ripple Δi are respectively expressed as

$$D_{y\_CCM}(t) = 1 - \frac{v_g(t)}{V_o} \tag{4.2}$$

$$\Delta i(t) = \frac{v_g(t) D_{y\_CCM}(t)}{L_b f_s} = \frac{v_g(t)}{L_b f_s}\left(1 - \frac{v_g(t)}{V_o}\right) \tag{4.3}$$

where, V_o is the output voltage, and f_s is the switching frequency.

Assuming that the efficiency is 100%, to ensure that the converter operates in CCM under the output power of P_o, the average value of the inductor current, i_{Lb_avg}, should satisfy

$$i_{Lb\_avg}(t) = \frac{\sqrt{2}P_o}{V_{in}}|\sin\omega_{in}t| \geq \frac{1}{2}\Delta i(t) \tag{4.4}$$

Substituting (4.1) into (4.3) and then combining with (4.4), we have

$$1 - \frac{\sqrt{2}V_{in}}{V_o}|\sin\omega_{in}t| \leq \frac{2P_o L_b f_s}{V_{in}^2} \tag{4.5}$$

As seen in (4.5), when $|\sin\omega_{in}t| = 0$, the left side has the maximum value. Thus, the output power in all-CCM is

$$P_{o\_all\text{-}CCM} \geq \frac{V_{in}^2}{2L_b f_s} \tag{4.6}$$

If the inequality (4.5) is not satisfied, the converter will be operated in DCM in a certain period in a half line cycle. Furthermore, if (4.5) is not satisfied at $T_{line}/4$, where $|\sin\omega_{in}t| = 1$, the converter will be always operated in DCM. Hence, substituting $|\sin\omega_{in}t| = 1$ into (4.5), the output power in all-DCM can be derived as

$$P_{o\_all\text{-}DCM} < \frac{V_{in}^2\left(V_o - \sqrt{2}V_{in}\right)}{2L_b f_s V_o} \tag{4.7}$$

The boost inductance L_b is usually designed according to λ_i, which is the ratio of the inductor current ripple Δi and the average inductor current i_{Lb_avg} under the minimum input voltage V_{in_min} and full-load P_{o_max} at $T_{line}/4$ [5]. According to (4.3) and (4.4), λ_i is derived as

$$\lambda_i = \frac{\Delta i}{i_{Lb\_avg}} = \frac{V_{in\_min}^2}{L_b P_{o\_max} f_s} \cdot \left(1 - \frac{\sqrt{2}V_{in\_min}}{V_o}\right) \tag{4.8}$$

According to (4.8), the boost inductance L_b is given by

$$L_b = \frac{V_{in\_min}^2}{\lambda_i P_{o\_max} f_s}\left(1 - \frac{\sqrt{2}V_{in\_min}}{V_o}\right) \tag{4.9}$$

Substituting (4.9) into (4.6) and (4.7), respectively, with the base of the rated output power, the normalized output power in all-CCM and all-DCM can be derived as

$$P^*_{o\_all\text{-}CCM} = \frac{P_{o\_all\text{-}CCM}}{P_{o\_max}} \geq \frac{\lambda_i V_{in}^2 V_o}{2V_{in\_min}^2\left(V_o - \sqrt{2}V_{in\_min}\right)} \tag{4.10}$$

Fig. 4.3 The regions for the all-CCM, part-CCM/DCM, and all-DCM

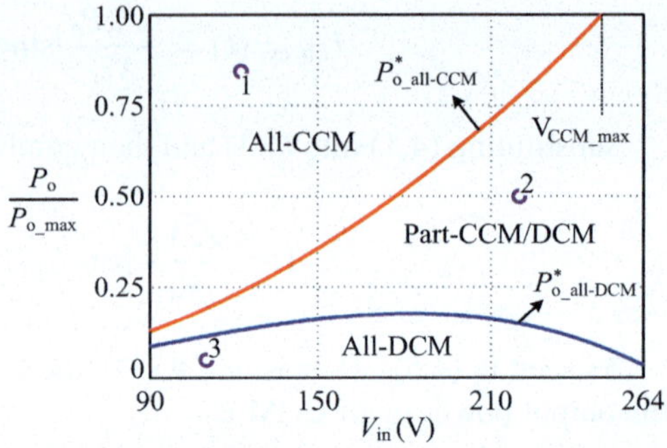

$$P^*_{o\_all\text{-}DCM} = \frac{P_{o\_all\text{-}DCM}}{P_{o\_max}} < \frac{\lambda_i V_{in}^2 \left(V_o - \sqrt{2} V_{in}\right)}{2 V_{in\_min}^2 \left(V_o - \sqrt{2} V_{in\_min}\right)} \tag{4.11}$$

According to (4.10) and (4.11), the regions of the input voltage and load condition for all-CCM, part-CCM/DCM and all-DCM are depicted in Fig. 4.3, where, $V_{in} =$ 90~264 V, $V_o = 385$ V, $\lambda_i = 20\%$, and $V_{all\text{-}CCM\_max}$ is the maximum input voltage for all-CCM under full-load. Based on (4.10), $V_{all\text{-}CCM\_max}$ is calculated by letting $P_{o\_all\text{-}CCM}$ equal to P_{o_max}, yields

$$V_{CCM\_max} = \sqrt{\frac{2 V_{in\_min}^2 \left(V_o - \sqrt{2} V_{in\_min}\right)}{\lambda_i V_o}} \tag{4.12}$$

As seen in Fig. 4.3, under the same input voltage V_{in}, when $P_o / P_{o\_max} > P^*_{o\_all\text{-}CCM}$, the converter operates in all-CCM; when $P^*_{o\_all\text{-}DCM} \leq P_o / P_{o\_max} \leq P^*_{o\_all\text{-}CCM}$, the converter operates in part-CCM/DCM; when $P_o / P_{o\_max} < P^*_{o\_all\text{-}DCM}$, the converter operates in all-DCM.

4.2 The Voltage Across the Power Switch Under Different Operation Modes

Take the positive half line cycle as an example, the switching modes of the boost PFC converter under three operating modes are illustrated in Fig. 4.4.

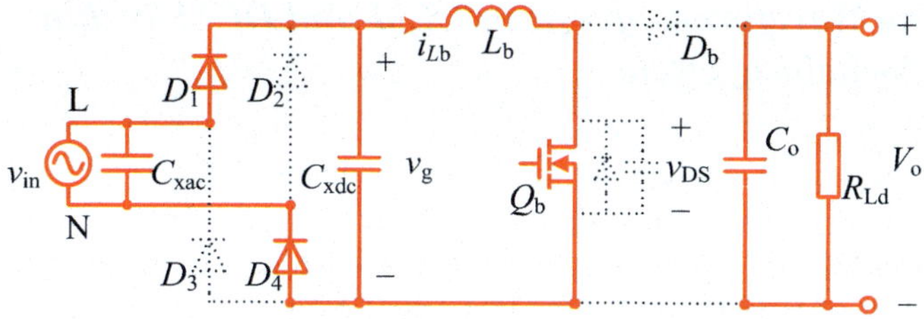

(a) Q_b is conducting

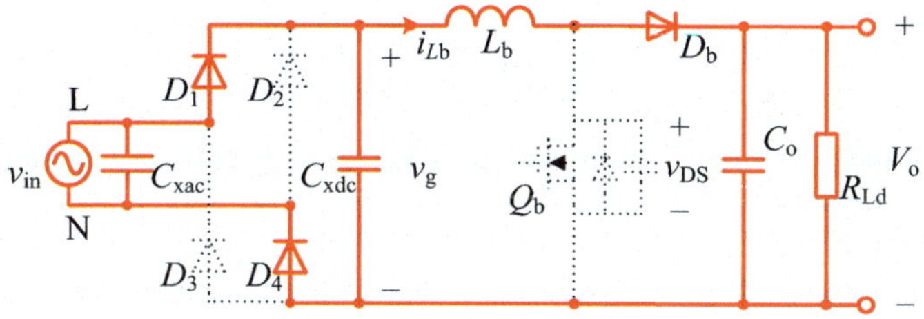

(b) Q_b is turned off

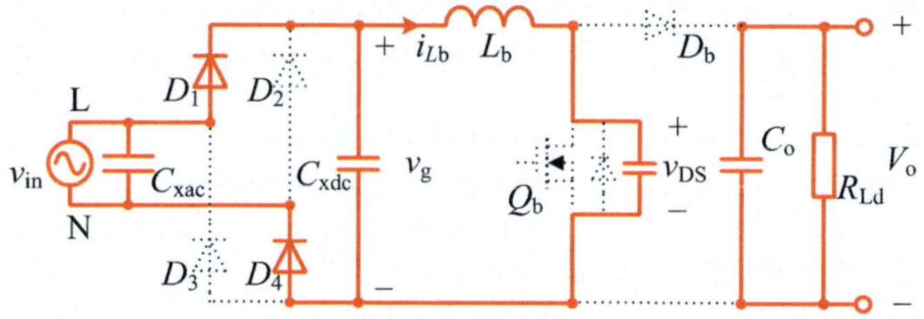

(c) Resonance between L_b and the capacitance of Q_b

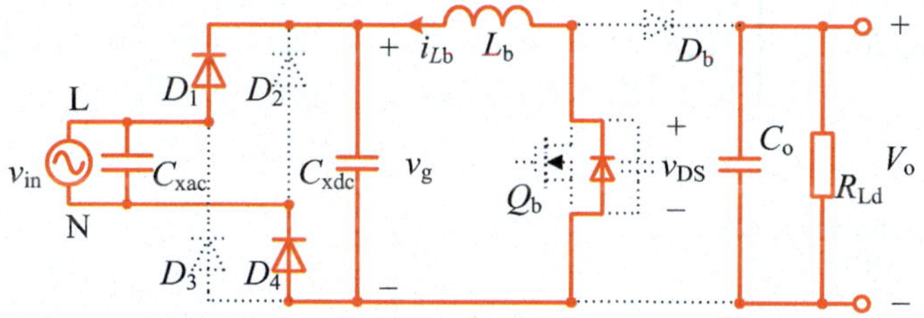

(d) The body diode of Q_b is conducting

Fig. 4.4 The switching modes of boost PFC converter operated in CCM and DCM

4.2.1 The Waveforms of v_{DS} in CCM and DCM Within a Switching Cycle

In CCM, when Q_b is conducting, its drain-to-source voltage v_{DS} is zero, and the inductor current i_{Lb} rises linearly, as shown in Fig. 4.4a. When Q_b is turned off, v_{DS} equals to V_o, and i_{Lb} decays linearly, as shown in Fig. 4.4b. Figure 4.5 gives the waveforms of v_{DS} and i_{Lb} for the boost PFC converter in CCM, and v_{DS} can be expressed as

$$v_{DS\_CCM}(t_s) = \begin{cases} 0 & t_s \in [0, D_{y\_CCM}T_s] \\ V_o & t_s \in (D_{y\_CCM}T_s, T_s] \end{cases} \tag{4.13}$$

In DCM, the possible waveforms of i_{Lb} and v_{DS} in a switching cycle is shown in Fig. 4.6, and the duty cycle is defined as D_{y_DCM}. In $[0, D_{y\_DCM}T_s]$, Q_b is conducting, and the switching mode is shown in Fig. 4.5a. In $[D_{y\_DCM}T_s, t_d]$, Q_b is turned off, and the switching mode is shown in Fig. 4.5b. At the time instant t_d, i_{Lb} decays to zero, and L_b resonates with the junction capacitor C_{DS} of Q_b (assuming that C_{DS} is far larger than the junction capacitor of diode D_b), as shown in Fig. 4.5c. If $2v_g \geq V_o$, the waveforms of i_{Lb} and v_{DS} are given in Fig. 4.6a, where the resonance between L_b and C_{DS} continues until Q_b is turned on, and the minimal value of v_{DS} is $2v_g - V_o$. If $2v_g < V_o$, the waveforms of i_{Lb} and v_{DS} are shown in Fig. 4.6b, where, v_{DS} drops to 0 at t_1, and the body diode of Q_b is conducting, as shown in Fig. 4.5d. As the voltage across L_b is v_g, i_{Lb} rises linearly. When i_{Lb} rises to zero at t_2, L_b and C_{DS} resonate again, and the switching mode is the same as Fig. 4.5c until Q_b is turned on. During $[t_2, T_s]$, the maximum of v_{DS} is $2v_g$.

According to Fig. 4.6, v_{DS} can be expressed as

$$v_{DS\_DCM\_a}(t_s) = \begin{cases} 0 & t_s \in [0, D_{y\_DCM}T_s] \\ V_o & t_s \in (D_{y\_DCM}T_s, t_d] \\ v_g + (V_o - v_g)\cos\omega_r(t_s - t_d) & t_s \in (t_d, T_s] \end{cases} \tag{4.14}$$

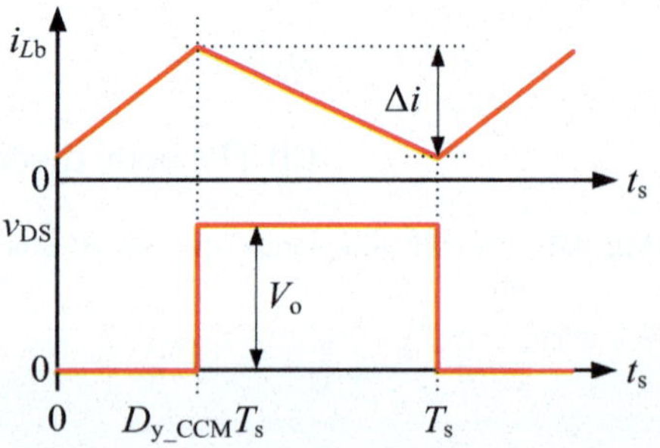

Fig. 4.5 Waveforms of i_{Lb} and v_{DS} in a switching cycle in CCM

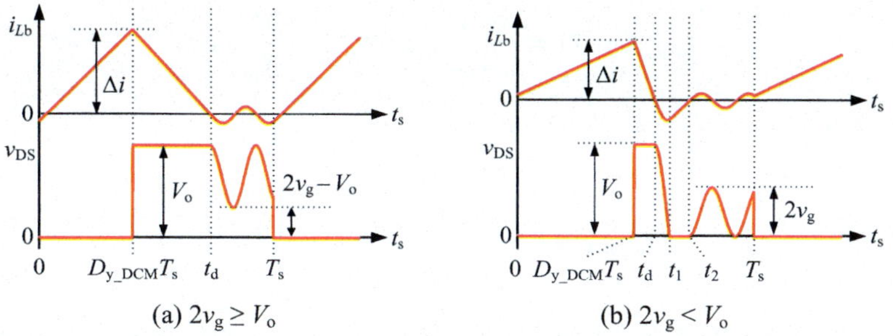

Fig. 4.6 Waveforms of i_{Lb} and v_{DS} in a switching cycle in DCM

$$v_{DS\_DCM\_b}(t_s) = \begin{cases} 0 & t_s \in [0, D_{y\_DCM}T_s] \text{ and } t_s \in (t_1, t_2] \\ V_o & t_s \in (D_{y\_DCM}T_s, t_d] \\ v_g + (V_o - v_g)\cos\omega_r(t_s - t_d) & t_s \in (t_d, t_1] \\ v_g - v_g \cos\omega_r(t_s - t_{o2}) & t_s \in (t_2, T_s] \end{cases} \tag{4.15}$$

where, ω_r is the resonant angular frequency of L_b and C_{DS}, and ω_r, t_d, t_1, and t_2 are given by

$$\omega_r = 1/\sqrt{L_b C_{DS}} \tag{4.16}$$

$$t_d = \frac{V_o}{V_o - v_g} D_{y\_DCM} T_s \tag{4.17}$$

$$t_1 = t_d + \frac{1}{\omega_r} \arccos\left(\frac{v_g}{v_g - V_o}\right) \tag{4.18}$$

$$t_2 = t_1 + \frac{1}{\omega_r v_g} \sqrt{(V_o - v_g)^2 - v_g^2} \tag{4.19}$$

In DCM, the peak value of the inductor current in a switching cycle is given by

$$i_{peak\_DCM}(t) = \frac{v_g(t) D_{y\_DCM}(t)}{L_b f_s} \tag{4.20}$$

Assuming that the resonance has negligible impact on the average inductor current, then the average inductor current can be obtained according to (4.17) and (4.20), expressed as

$$i_{avg\_DCM}(t) = \frac{t_d}{2T_s} i_{peak\_DCM}(t) = \frac{1}{2} \frac{V_o}{V_o - v_g(t)} \frac{v_g(t)}{L_b f_s} D_{y\_DCM}^2(t) \tag{4.21}$$

With the average-current control, the average input current is given by

$$i_{avg\_DCM}(t) = \frac{\sqrt{2}P_o}{V_{in}}|\sin\omega_{in}t| = \frac{P_o}{V_{in}^2}v_g(t) \tag{4.22}$$

According to (4.21) and (4.22), D_{y_DCM} is derived as

$$D_{y\_DCM}(t) = \sqrt{\frac{2P_oL_bf_s(V_o - v_g(t))}{V_{in}^2V_o}} \tag{4.23}$$

In part-CCM/DCM, the boost PFC converter will transit from DCM to CCM within [0, $T_{line}/4$], and the time instant of transition is defined as t_b. According to (4.5), t_b is derived as

$$t_b = \frac{1}{\omega_{in}}\arcsin\left(\frac{V_o}{\sqrt{2}V_{in}} - \frac{\sqrt{2}P_oV_oL_bf_s}{V_{in}^3}\right), \quad t_b \in [0, T_{line}/4] \tag{4.24}$$

4.2.2 The Waveforms of Duty Cycle and v_{DS} in a Half Line Cycle

Referring to (4.2), (4.23) and (4.24), the duty cycle in a half line cycle for all-CCM, part-CCM/DCM and all-DCM under the same input voltage is given in Fig. 4.7. Where, in all-CCM and all-DCM, the duty cycle within [0, $T_{line}/4$] is D_{y_CCM} and D_{y_DCM}, respectively; in part-CCM/DCM, the duty cycle within [0, t_b] and [t_b, $T_{line}/4$] is D_{y_DCM} and D_{y_CCM}, respectively. According to (4.2) and (4.23), D_{y_DCM} is always less than D_{y_CCM} in a half line cycle under the same input rectified voltage. The lighter the load is, the less the duty cycle D_{y_DCM} will be.

Referring to Fig. 4.6, i_{Lb} and v_{DS} have two patterns under DCM, so $t_{2vg=Vo}$ is defined as the time instant when $2v_g = V_o$ within [0, $T_{line}/4$]. Figure 4.8 shows the waveforms of v_g and v_{DS} under different operation modes. As seen, these curves are symmetrical to $T_{line}/4$.

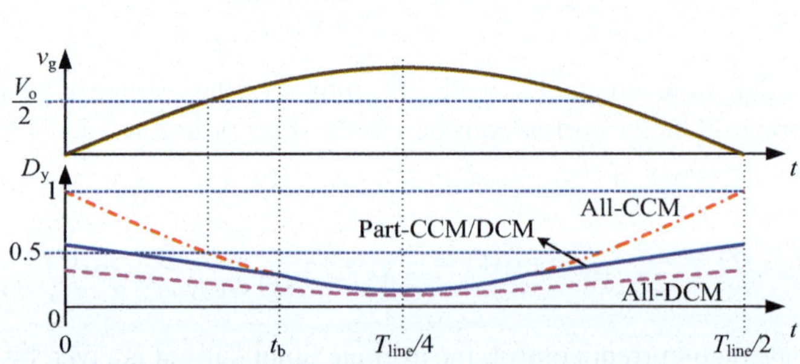

Fig. 4.7 The curves of the duty cycle under different operation modes

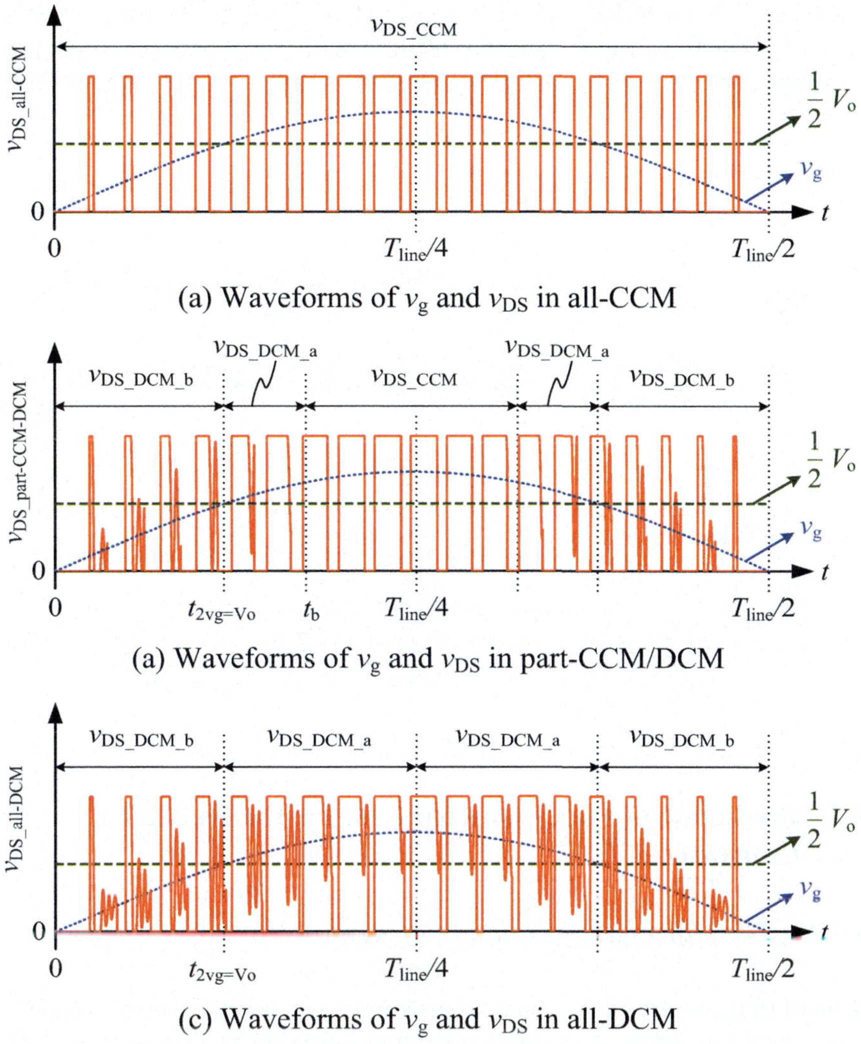

Fig. 4.8 Waveforms of v_g and v_{DS} in all operation modes

In all-CCM, Fig. 4.8a shows $v_{DS\_all\text{-}CCM}$ in a half line cycle. As seen, $v_{DS\_all\text{-}CCM}$ is v_{DS_CCM} for each switching cycle, and its duty cycle varies in a half line cycle, yields

$$v_{DS\_all\text{-}CCM}(t, t_s) = \begin{cases} 0 & t_s \in [0,\ D_{y\_CCM}(t)T_s] \\ V_o & t_s \in (D_{y\_CCM}(t)T_s,\ T_s] \end{cases} \quad (0 \le t \le T_{line}/2) \tag{4.25}$$

In part-CCM/DCM, Fig. 4.8b shows $v_{DS\_part\text{-}CCM/DCM}$ in a half line cycle when $t_b \geq t_{2vg=Vo}$. Within [0, $t_{2vg=Vo}$], [$t_{2vg=Vo}$, t_b] and [t_b, $T_{line}/4$], $v_{DS\_part\text{-}CCM/DCM}$ is $v_{DS\_DCM\_b}$, $v_{DS\_DCM\_a}$ and v_{DS_CCM} for each switching cycle, respectively, and the duty cycle is D_{y_DCM} and D_{y_CCM} within [0, t_b] and [t_b, $T_{line}/4$], respectively. For the case when $t_b < t_{2vg=Vo}$, $v_{DS\_part\text{-}CCM/DCM}$ is $v_{DS\_DCM\_b}$ within [0, t_b], and $v_{DS\_DCM\_a}$ will not appear. The expressions for $v_{DS\_part\text{-}CCM/DCM}$ is

$$v_{DS\_part\text{-}CCM/DCM}(t, t_s) = \left| \begin{array}{ll} v_{DS\_DCM\_b}(t_s) & \left(0 \leq t \leq t_{2vg=Vo}\right) \\ v_{DS\_DCM\_a}(t_s) & \left(t_{2vg=Vo} \leq t \leq t_b\right) \\ v_{DS\_CCM}(t_s) & \left(t_b \leq t \leq T_{line}/2 - t_b\right) \\ v_{DS\_DCM\_a}(t_s) & \left(T_{line}/2 - t_b \leq t \leq T_{line}/2 - t_{2vg=Vo}\right) \\ v_{DS\_DCM\_b}(t_s) & \left(T_{line}/2 - t_{2vg=Vo} \leq t \leq T_{line}/2\right) \end{array} \right. \tag{4.26}$$

In all-DCM, Fig. 4.8c shows $v_{DS\_all\text{-}DCM}$ in a half line cycle. Within [0, $t_{2vg=Vo}$] and [$t_{2vg=Vo}$, $T_{line}/4$], $v_{DS\_all\text{-}DCM}$ is $v_{DS\_DCM\_b}$ and $v_{DS\_DCM\_a}$ for each switching cycle, respectively, and the duty cycle is D_{y_DCM} within [0, $T_{line}/4$], yields

$$v_{DS\_all\text{-}DCM}(t, t_s) = \left| \begin{array}{ll} v_{DS\_DCM\_b}(t_s) & \left(0 \leq t \leq t_{2vg=Vo}\right) \\ v_{DS\_DCM\_a}(t_s) & \left(t_{2vg=Vo} \leq t \leq T_{line}/2 - t_{2vg=Vo}\right) \\ v_{DS\_DCM\_b}(t_s) & \left(T_{line}/2 - t_{2vg=Vo} \leq t \leq T_{line}/2\right) \end{array} \right. \tag{4.27}$$

4.3 Prediction of the Conducted EMI for the Boost PFC Converter

4.3.1 The Harmonic Amplitude of v_{DS}

In boost PFC converters, v_{DS} can be treated as the voltage source that generates the conducted EMI, and the short-time Fourier transform (STFT) is adopted to calculate the harmonic amplitude of v_{DS} in a half line cycle [6]. By treating the waveform in a switching period as the window for calculation, the harmonic amplitudes of v_{DS} are expressed as

$$V_{DS\_k}(t) = \frac{2}{T_s}\left| \int_0^{T_s} v_{DS}(t, t_s) \cdot e^{-j2k\pi f_s t_s} dt_s \right| \tag{4.28}$$

where, k is the harmonic order. Since the switching frequency of average-current-controlled boost PFC converter is constant, the conducted electromagnetic interference (EMI) is mainly distributed at the multiples of the switching frequency in the spectrum.

In all-CCM, v_{DS} is $v_{DS\_all\text{-}CCM}$ within $[0, T_{line}/4]$. By substituting (4.25) into (4.28), the harmonic amplitude of $v_{DS\_all\text{-}CCM}$ in all-CCM is

$$V_{DS\_k}(t) = \frac{2}{T_s}\left|\int_0^{T_s} v_{DS\_all\text{-}CCM}(t, t_s) \cdot e^{-j2k\pi f_s t_s} dt_s\right| = \frac{2V_o}{k\pi}\left|\sin\left(k\pi\frac{v_g(t)}{V_o}\right)\right|, \quad \text{(all-CCM)} \tag{4.29}$$

Referring to (4.29), in all-CCM, the harmonic amplitude of v_{DS} is only related to the input voltage v_g and V_o, and is irrelevant to the load condition.

In part-CCM/DCM and all-DCM, the oscillation caused by the boost inductor and the power switch junction capacitor affects harmonic amplitude of v_{DS} in a half line cycle. Substituting (4.26) and (4.27) into (4.28), the harmonic amplitudes of v_{DS} in part-CCM/DCM and all-DCM can be calculated, which are expressed as

$$V_{DS\_k}(t) = \frac{2}{T_s}\left|\int_0^{T_s} v_{DS\_part\text{-}CCM/DCM}(t, t_s) \cdot e^{-j2k\pi f_s t_s} dt_s\right|, \quad \text{(part-CCM/DCM)} \tag{4.30}$$

$$V_{DS\_k}(t) = \frac{2}{T_s}\left|\int_0^{T_s} v_{DS\_all\text{-}DCM}(t, t_s) \cdot e^{-j2k\pi f_s t_s} dt_s\right|, \quad \text{(all-DCM)} \tag{4.31}$$

In part-CCM/DCM and all-DCM, since $D_{y\_DCM}(t)$ is related to v_g, V_o and P_o, the harmonic amplitude of v_{DS} is also related to v_g, V_o and P_o.

4.3.2 Characteristics of the Conducted EMI of the Boost PFC Converter

In Chap. 3, it is pointed out that the common-mode (CM) and differential-mode (DM) noises V_{CM_k} and V_{DM_k} of the boost PFC converter is related to the spectrum of v_{DS}. According to (3.8) and (3.9), V_{CM_k} and V_{DM_k} can be calculated by multiplying V_{DS_k} with the CM and DM noise voltage gains $|\text{CMTG}(kf_s)|$ and $|\text{DMTG}(kf_s)|$ of the boost PFC converter, respectively, expressed as

$$\begin{cases} V_{CM\_k} = |\text{CMTG}(kf_s)| \cdot V_{DS\_k} = \pi k f_s C_p R_{LN} \cdot V_{DS\_k} \\ V_{DM\_k} = |\text{DMTG}(kf_s)| \cdot V_{DS\_k} = \frac{1}{8\pi^2 (kf_s)^2 L_b C_{x1}} \cdot V_{DS\_k} \end{cases} \tag{4.32}$$

Based on the principle of EMI receiver in Fig. 2.7, the CM and DM noises of the boost PFC converter are sent to the equivalent intermediate-frequency (IF) filter. When the center frequency f_{IF} equals to kf_{s}, the IF filter has unity gain to the harmonic, and the envelope signal $V_{\text{CM\_}k\text{\_env}}$ ($V_{\text{DM\_}k\text{\_env}}$) equals to $V_{\text{CM\_}k}$ ($V_{\text{DM\_}k}$) exactly. Therefore, the peak (PK), quasi-peak (QP) and average (AV) values of the CM and DM noise spectrum are related to the time-domain characteristics of $V_{\text{CM\_}k}$ and $V_{\text{DM\_}k}$, respectively. According to (4.32), when the main circuit parameters are determined, |CMTG(kf_{s})| and |DMTG(kf_{s})| do not change over time. Therefore, the PK, QP and AV values of the CM and DM noise are expressed as

$$\begin{cases} V_{\text{CM\_}k\text{\_PK/QP/AV}} = \pi k f_{\text{s}} C_{\text{p}} R_{\text{LN}} \cdot V_{\text{DS\_}k\text{\_ PK/QP/AV}} \\ V_{\text{DM\_}k\text{\_PK/QP/AV}} = \frac{1}{8\pi^2 (kf_{\text{s}})^2 L_{\text{b}} C_{\text{x1}}} \cdot V_{\text{DS\_}k\text{\_ PK/QP/AV}} \end{cases} \tag{4.33}$$

where, $V_{\text{DS\_}k\text{\_PK/QP/AV}}$ is the PK, QP and AV values of $V_{\text{DS\_}k}$ in a half line cycle. $V_{\text{DS\_}k\text{\_PK}}$ and $V_{\text{DS\_}k\text{\_AV}}$ are given by

$$V_{\text{DS\_}k\text{\_PK}} = \max\{V_{\text{DS\_}k}(t)\} \quad t \in [0, T_{\text{line}}/2] \tag{4.34}$$

$$V_{\text{DS\_}k\text{\_AV}} = \frac{2}{T_{\text{line}}} \int_0^{\frac{T_{\text{line}}}{2}} V_{\text{DS\_}k}(t)\, dt \tag{4.35}$$

The QP value $V_{\text{DS\_}k\text{\_QP}}$ is usually calculated by numerical method [7], and there is no explicit expression for $V_{\text{DS\_}k\text{\_QP}}$. Furthermore, the QP value is quite close to the PK value, so the QP value can be approximated by the PK value and it is not shown here.

Based on (4.30), the characteristic of the CM and DM noises of the boost PFC converter are related to $V_{\text{DS\_}k\text{\_PK/AV}}$. Since v_{DS} has different waveforms under different input voltage and load conditions, the characteristic of $V_{\text{DS\_}k\text{\_PK/AV}}$ will be analyzed in all-CCM, part-CCM/DCM and all-DCM to predict the conducted EMI spectrum of boost PFC converter.

1. Characteristics of $V_{\text{DS\_}k\text{\_PK/AV}}$ in All-CCM

Based on the harmonic amplitude of v_{DS} in all-CCM in (4.29), $V_{\text{DS\_}k\text{\_CCM}}$ reaches its maximum when $k\pi v_{\text{g}}(t)/V_{\text{o}} = \pi/2$. Referring to (4.1), when $V_{\text{in}} < V_{\text{o}}/(2\sqrt{2})$, $\pi v_{\text{g}}(t)/V_{\text{o}}$ is always less than $\pi/2$, so the PK value of the fundamental harmonic $V_{\text{DS\_1\_PK\_CCM}}$ is always less than $2V_{\text{o}}/\pi$; when $V_{\text{o}}/(2\sqrt{2}) \le V_{\text{in}} \le V_{\text{CCM\_max}}$, $\pi v_{\text{g}}(t)/V_{\text{o}} = \pi/2$ at certain time instant in a half line cycle, and $V_{\text{DS\_1\_PK\_CCM}}$ equals to $2V_{\text{o}}/\pi$. When $k \ge 2$, there always exist $k\pi v_{\text{g}}(t)/V_{\text{o}} = \pi/2$ at a certain instant in a half line cycle, this means

$$\begin{cases} V_{DS\_1\_PK\_CCM} = \begin{vmatrix} \frac{2V_o}{\pi}\sin\left(\pi\frac{\sqrt{2}V_{in}}{V_o}\right), & \left(V_{in} < V_o/2\sqrt{2}\right) \\ \frac{2V_o}{\pi}, & \left(V_o/2\sqrt{2} \le V_{in} \le V_{CCM\_max}\right) \end{vmatrix} \\ V_{DS\_k\_PK\_CCM} = \frac{2V_o}{k\pi}, \quad (k \ge 2,\ V_{in} \le V_{CCM\_max}) \end{cases} \tag{4.36}$$

As seen, the maximum of PK values for the kth harmonic amplitude of v_{DS} in all-CCM is $2V_o/(k\pi)$, and it decays with a slope of − 20 dB/dec against the frequency because $2V_o/(k\pi)$ is inversely proportional to k. As described in Sect. 3.1.2, $|\mathrm{CMTG}(kf_s)|$ increases with a slope of 20 dB/dec, whereas $|\mathrm{DMTG}(kf_s)|$ decays with a slope of − 40 dB/dec. According to (4.33), in all-CCM, the attenuation slopes of the largest PK values of the CM and DM noise are 0 dB/dec and − 60 dB/dec, respectively.

Substituting (4.1) and (4.29) into (4.36), the AV value of the harmonic amplitude for v_{DS} in all-CCM are expressed as

$$V_{DS\_k\_AV\_CCM} = \frac{2V_o}{k\pi^2}\int_0^{\pi}\left|\sin\left(\sqrt{2}\pi\frac{kV_{in}}{V_o}\sin\omega_{in}t\right)\right|d\omega_{in}t = \frac{2V_o}{k\pi}F_{AV}(kV_{in}/V_o) \tag{4.37}$$

where, $F_{AV}(kV_{in}/V_o)$ is the characteristic function for calculating the AV value, expressed as

$$F_{AV}(kV_{in}/V_o) = \frac{1}{\pi}\int_0^{\pi}\left|\sin\left(\sqrt{2}\pi\frac{kV_{in}}{V_o}\sin\omega_{in}t\right)\right|d\omega_{in}t \tag{4.38}$$

According to (4.38), $F_{AV}(kV_{in}/V_o)$ is plotted in Fig. 4.9. Table 4.1 lists the local maximum of $F_{AV}(kV_{in}/V_o)$, and kV_{in}/V_o lies in $[kV_{in\_min}/V_o, kV_{CCM\_max}/V_o]$. For the kth harmonic, the input voltage for achieving the largest AV value can be found by comparing $F_{AV}(kV_{in}/V_o)$ at kV_{in_min}/V_o, kV_{CCM_max}/V_o and the local maximum in $[kV_{in\_min}/V_o, kV_{CCM\_max}/V_o]$. For example, with V_{in} = 90~264 V, V_o = 385 V and $\lambda_i = 0.18$ in the experiment, V_{CCM_max} is calculated as 245.5 V according to (4.12). When $k = 1$, based on Fig. 4.9 and Table 4.1, V_{in}/V_o is within [0.234, 0.637]. When $kV_{in}/V_o = 0.45$ (V_{in} = 173 V), $F_{AV}(kV_{in}/V_o)$ reaches the local maximum, which is the largest AV value for fundamental harmonic of v_{DS}.

According to (4.37), the AV values for each harmonic of v_{DS} under different input voltage in all-CCM are depicted in Fig. 4.10. By comparing Figs. 4.9 and 4.10, the variation of AV value for the kth harmonic with respect to the input voltage V_{in} is identical to the characteristics of the function $F_{AV}(kV_{in}/V_o)$ in $[kV_{in\_min}/V_o, kV_{CCM\_max}/V_o]$. Furthermore, the largest AV values of the second harmonic appear at V_{in_min} = 90 V and that of the third harmonic appear at $3V_{in}/V_o = 1.17$, i.e., V_{in} = 150 V, and so on.

Referring to Fig. 4.9, it is worth noting that the value of $F_{AV}(kV_{in}/V_o)$ after V_{in_min}/V_o is always larger than 0.49 (−6.2 dB). Therefore, according to (4.33) and (4.34),

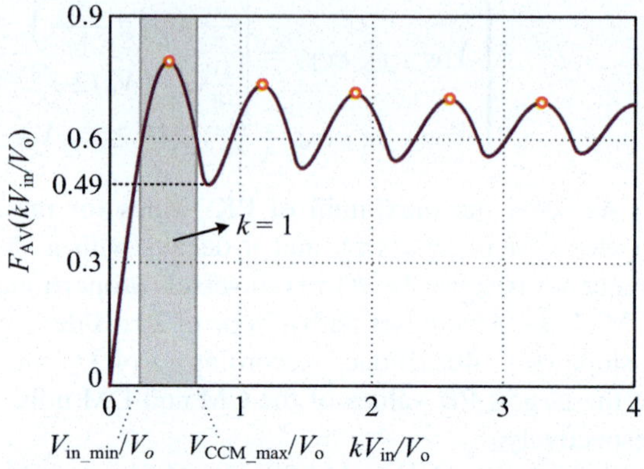

Fig. 4.9 Numerical calculated curve of the function $F_{AV}(kV_{in}/V_o)$

Table 4.1 Local Maximum of the Function $F_{AV}(kV_{in}/V_o)$

kV_{in}/V_o	0.45	1.17	1.87	2.58	3.28	...
$F_{AV}\ (kV_{in}/V_o)$	0.79	0.73	0.71	0.70	0.69	...

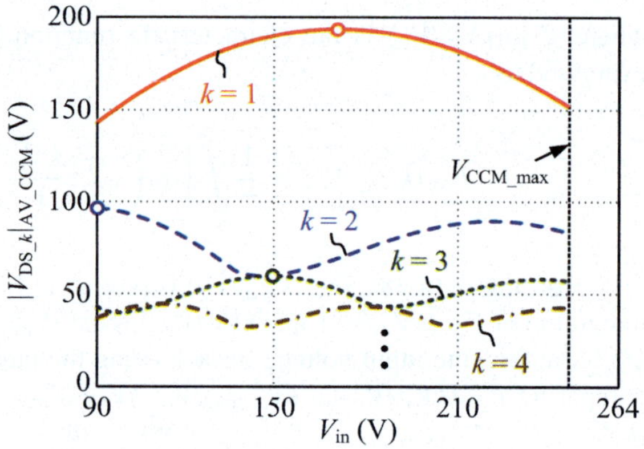

Fig. 4.10 Average values of $|v_{DS\_k}(t)|$ in all-CCM

the PK values are higher than the AV values by less than 6.2 dB. Recalling that in the EN 55022 standard, the QP limit is 10 dB higher than the AV limit for the equipments in Class B. As a consequence, the AV values of the boost PFC converter under all-CCM are critical for the EMI filter design, which will be further elaborated in Sect. 4.4.

2. Characteristics of $V_{DS\_k\_AV}$ in Part-CCM/DCM and All-DCM

According to (4.30) and (4.31), the expressions of $V_{DS\_k\_part\text{-}CCM/DCM}$ and $V_{DS\_k\_all\text{-}DCM}$ are quite complex to be analyzed. Therefore, the numerical method is adopted to reveal the characteristics of the conducted EMI in part-CCM/DCM and all-DCM.

According to (4.29), (4.30) and (4.31), the AV values of the fundamental, second, and third harmonic amplitudes of v_{DS} are shown in Fig. 4.11, where, the operation modes are divided by the dashed lines. The output voltage V_o is 385 V, and the resonant frequency f_r is $4.2f_s$, which is measured in the experiment.

In Fig. 4.11a, the AV value for the fundamental harmonic is independent of the load condition in all-CCM, and it is the largest when $V_{in}/V_o = 0.45$. Moreover, the AV value of V_{DS_1} in part-CCM/DCM under $V_{in}/V_o = 0.45$ is the largest under all working conditions. In all-DCM, V_{DS_1} is very low under the maximum input voltage and the lightest load. Here, 10% load is taken as the lightest load for consideration, since the converter may operate in bang-bang mode at light load and its conducted EMI can be even lower. As shown in Fig. 4.11b, for the second harmonic, the worst case may appear under high V_{in} and low P_o in part-CCM/DCM or all-DCM. For the third harmonic, the worst case also appears under high V_{in} and low P_o in part-CCM/DCM or all-DCM, as shown in Fig. 4.11c.

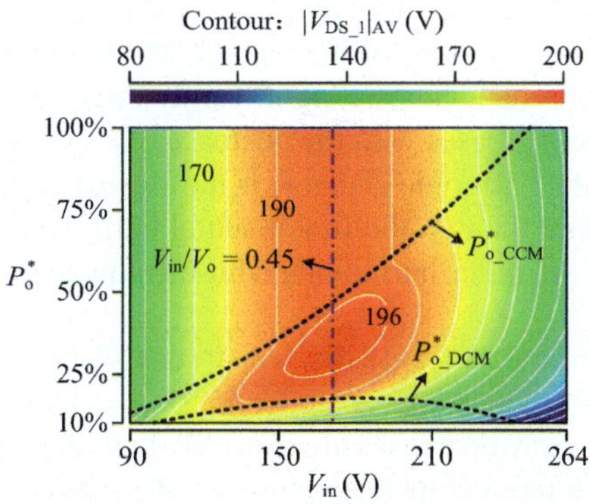

(a) Fundamental harmonic

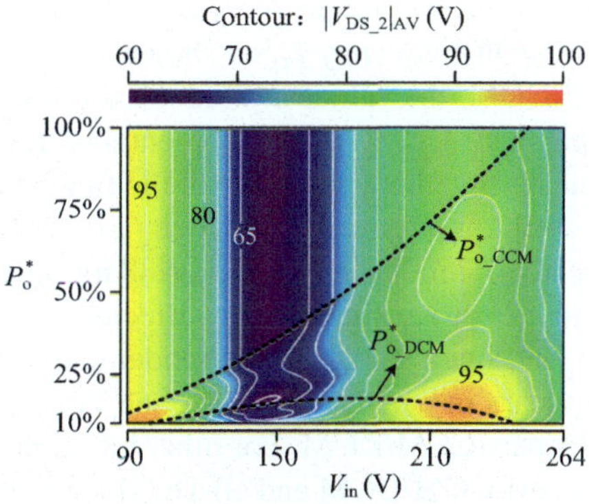

(b) Second harmonic

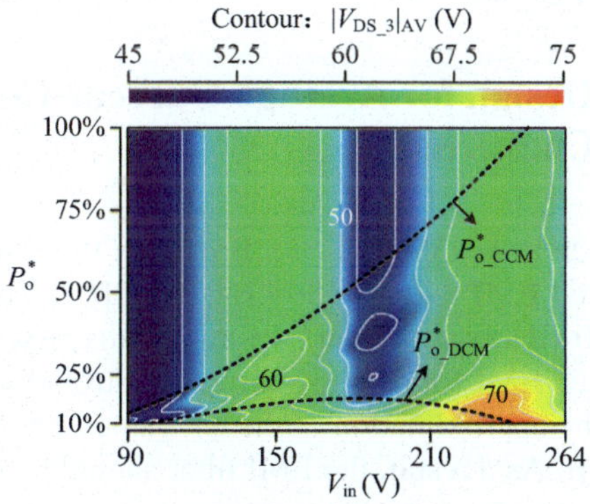

(c) Third harmonic

Fig. 4.11 Average value of $|v_{DS\_k}|$

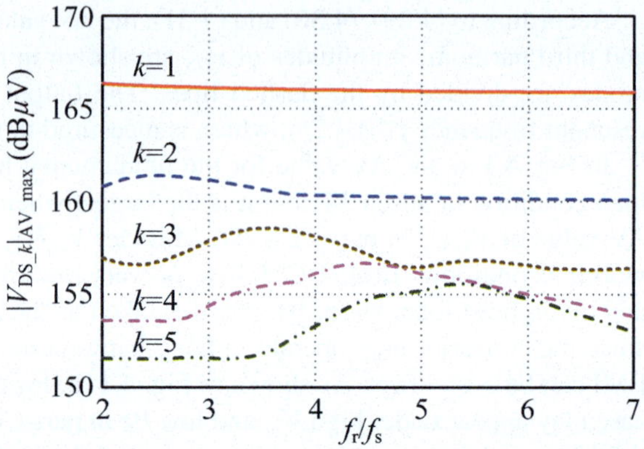

Fig. 4.12 Effect of the oscillation frequency on the conducted EMI

According to (4.29), (4.30), (4.31) and (4.37), the maximum AV value of each harmonic amplitude of v_{DS} in a half line cycle can be calculated with different resonant frequency f_r, as shown in Fig. 4.12. As seen, in the part-CCM/DCM and all-DCM, the values around the oscillation frequencies are significantly larger than that in all-CCM. The fundamental harmonic increases with the decrease of f_r, which means that the worst case for the fundamental harmonic is that f_r is the closest to the switching frequency f_s. When f_r is far higher than f_s, the fundamental harmonic is scarcely influenced by the oscillation, but the other harmonics whose frequency is close to f_r are significantly affected.

In conclusion, for the fundamental harmonic of v_{DS}, the worst case distinctly appears in part-CCM/DCM, and largest AV values appear under $V_{in}/V_o = 0.45$ in the part-CCM/DCM. For other harmonics, the worst cases are most likely to appear in part-CCM/DCM and all-DCM under high input voltage and light-load conditions.

4.4 EMI Filter Design for Average-Current-Controlled Boost PFC Converter

The EMI standard such as EN55022 Class B defines the limit for the QP and AV values of the conducted EMI within [150 kHz, 30 MHz]. The AV limit defined in EN55022 Class B is 10 dB lower than the QP limit. As discussed in Sect. 4.3.2, the PK values of the conducted EMI for average-current-controlled boost PFC converters, which are treated as the approximation of the QP values, are higher than AV values by no more than 10 dB. As a result, the EMI filter should be designed based on the largest AV values.

The attenuation requirement for the EMI filter is equal to the original CM and DM noises substracted by the limit, where, the original CM and DM noises are related to |CMTG(kf_s)|, |DMTG(kf_s) and the spectrum of v_{DS}. Referring to Sect. 4.3.2, the

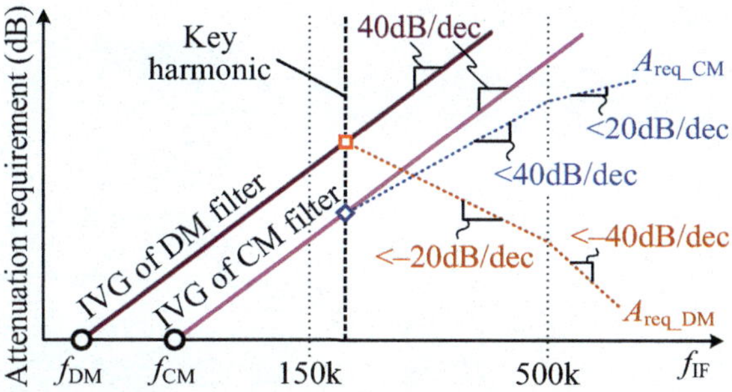

Fig. 4.13 Critical harmonics for EMI filter design

slopes of $|\mathrm{CMTG}(kf_s)|$ and $|\mathrm{DMTG}(kf_s)|$ are 20 dB/dec and − 40 dB/dec, respectively. According to (4.32), considering that the harmonics of v_{DS} attenuates with the frequency, the attenuation slopes of the CM and DM noise are lower than 20 dB/dec and − 40 dB/dec. Besides, the slopes of the limit within [150 kHz, 500 kHz] and [500 kHz, 1 MHz] are − 20 dB/dec and 0 dB/dec, respectively. As a result, the slope of the insert voltage gain (IVG) of the CM and DM filter in Fig. 3.11, which is 40 dB/dec, is larger than that of the attenuation requirement, as shown in Fig. 4.13. Therefore, the first harmonic after 150 kHz is the critical harmonic for the EMI filter design. When f_s is higher than 150 kHz, the fundamental harmonic is the critical harmonic; when f_s is within [75 kHz, 150 kHz], the second harmonic is the critical harmonic; and so on.

In addition, as revealed in Sect. 4.3.2, when the oscillation frequency is far higher than the switching frequency, the harmonic at the oscillation frequency in the part-CCM/DCM and all-DCM is significantly larger than that in the all-CCM. Considering the nonideality of the filter components at high frequencies, care must be taken with the harmonic at the oscillation frequency under light-load condition, when the designed filter is inserted. If it is necessary to suppress the EMI around oscillation frequency, methods of canceling the parasitic parameters in the EMI filter can be applied.

The EMI filter design for the average-current-controlled boost PFC converter can be summarized as follows.

(1) Predict the largest AV values of the first harmonic of the CM and DM noise after 150 kHz. If the critical harmonic is the fundamental harmonic, the largest AV values can be found in part-CCM/DCM under the condition $V_{in}/V_o = 0.45$. If the critical harmonic is the second or third harmonic, the largest AV values of the harmonic can be found in part-CCM/DCM or all-DCM under high input voltage and light-load conditions.
(2) Calculate the attenuation requirements for the critical harmonic of the CM and DM noise, and design the parameters of the filter components.
(3) If the oscillation frequency of boost inductor and the junction capacitor is far higher than the switching frequency, it is necessary to check the harmonics around the oscillation frequency under light-load condition.

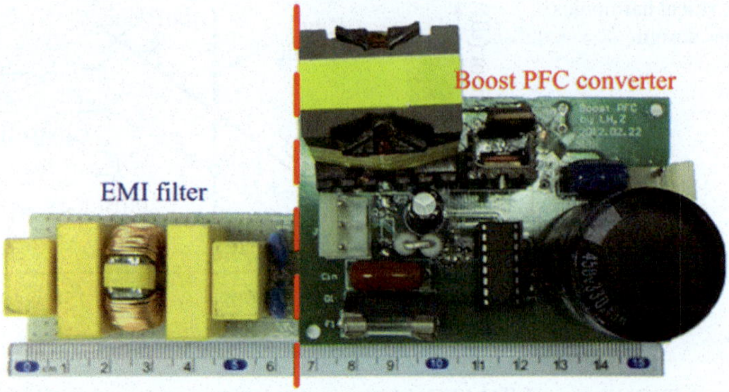

Fig. 4.14 Photograph of the 300-W prototype of the boost PFC converter

4.5 Experimental Results

4.5.1 Specifications of the Prototype

In order to verify the input voltage and load conditions for the worst spectrum of the conducted EMI, and design the EMI filter for the average-current-controlled boost PFC converter, a 300-W boost PFC converter prototype, as shown in Fig. 4.14, has been fabricated with the following specifications:

- Input voltage V_{in}: 90~264 V/50 Hz;
- Output voltage V_{o}: 385 V;
- Switching frequency f_{s}: 175 kHz (In order to acquire the fundamental harmonic in the conducted EMI, the switching frequency is intentionally designed to be higher than 150 kHz);
- Boost inductor L_{b}: 580 μH (L_{b} is designed according to (4.9), where $\lambda_i = 0.18$);
- Control IC: UC3854 (Unitrode).

4.5.2 Experimental Results

Figure 4.15 gives the waveforms for the input rectified voltage, inductor current and the drain-to-source voltage of the power switch under 220-V input voltage. As seen, in a half line cycle, the boost PFC converter is operated in all-CCM, part-CCM/DCM and all-DCM under full-load, 20% load and 10% load condition, respectively.

Figure 4.16 shows the waveforms of the inductor current $i_{L\mathrm{b}}$ and drain-to-source voltage v_{DS} of the power switch at different time intervals under 220-V input voltage and 20% load condition. As shown in Fig. 4.16a, when operated in CCM, $i_{L\mathrm{b}}$

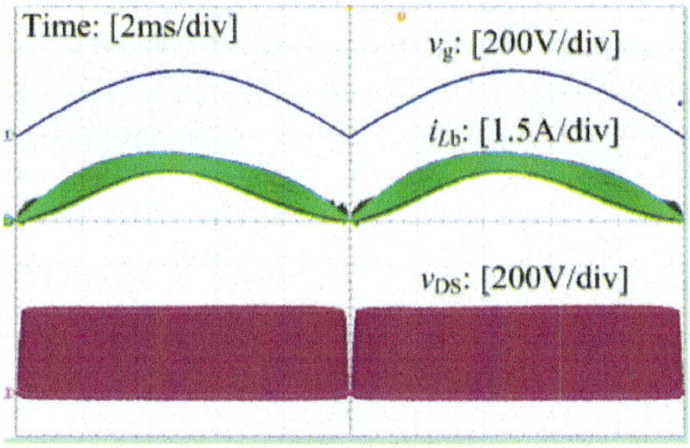

(a) All-CCM (full-load)

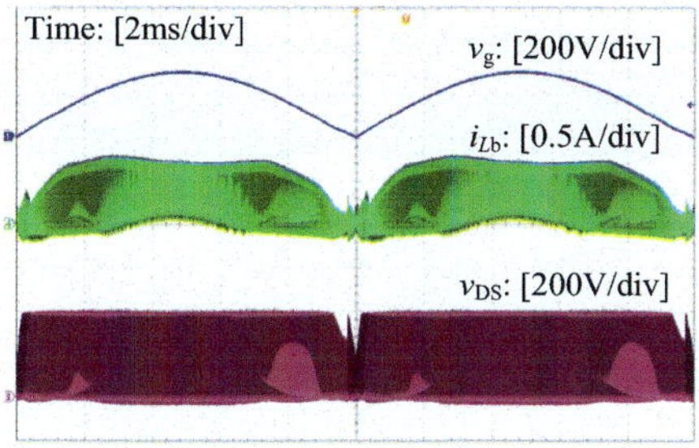

(b) Part-CCM/DCM (20% load)

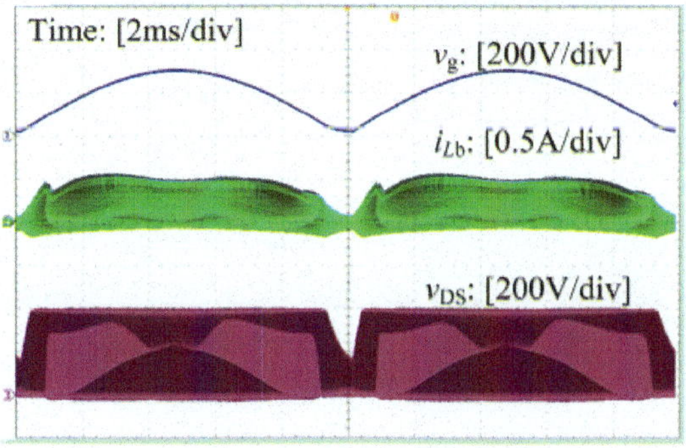

(c) All-DCM (10% load)

Fig. 4.15 Three operating modes of the boost PFC converter with 220-V input voltage

increases and decreases linearly with the power switch turned on and off, respectively. Figure 4.16b gives the waveforms of i_{Lb} and v_{DS} in DCM when $v_g > V_o/2$. As seen, i_{Lb} drops linearly with the power switch turned off, and the resonance in v_{DS} can be observed when $i_{Lb} < 0$. Furthermore, v_{DS} is always above zero during the resonance until the power switch is turned on. Figure 4.16c shows the waveforms of i_{Lb} and v_{DS} in DCM when $v_g < V_o/2$. As seen, there is a certain interval for v_{DS} being zero when the power switch is turned off, which verifies the previous analysis.

Figure 4.17 gives the AV values for the original CM and DM noises at 175 kHz (the fundamental harmonic), which are both measured under different input and load conditions, respectively. In addition, the dashed lines are adopted to divide the operation modes. Figure 4.18 shows the PK and AV values of the original CM and DM noise measured at 175 kHz. According to Figs. 4.17 and 4.18, the measured conducted EMI in all-CCM is independent of load and only related with the input voltage, as it is revealed in Sect. 4.2.2. The largest AV values of the CM and DM noise at 175 kHz appears in the part-CCM/DCM under 170 and 180 V, which are close to $V_{in}/V_o = 0.45$, thus verifying the theoretical analysis.

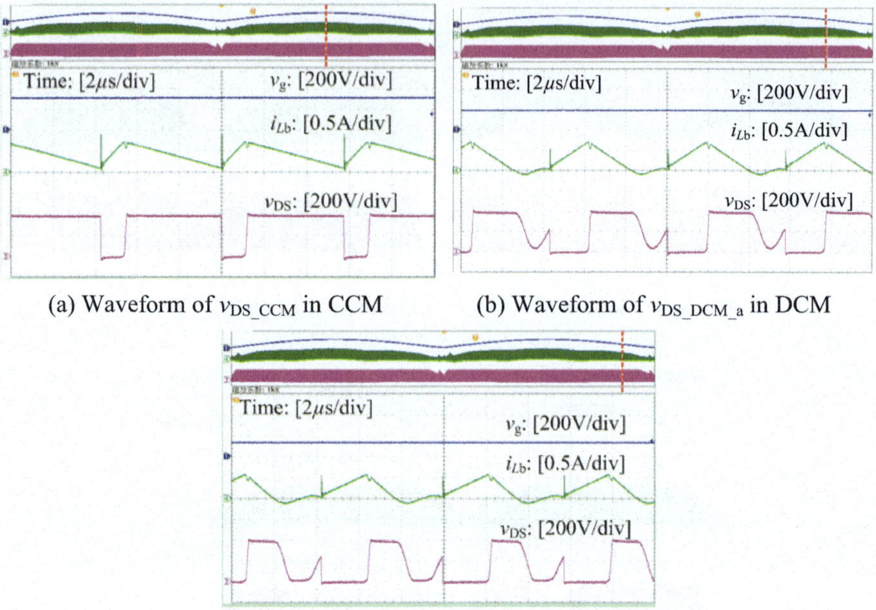

(a) Waveform of v_{DS_CCM} in CCM (b) Waveform of $v_{DS\_DCM\_a}$ in DCM

(c) Related waveform of $v_{DS\_DCM\_b}$ in DCM

Fig. 4.16 The related waveforms in part-CCM/DCM

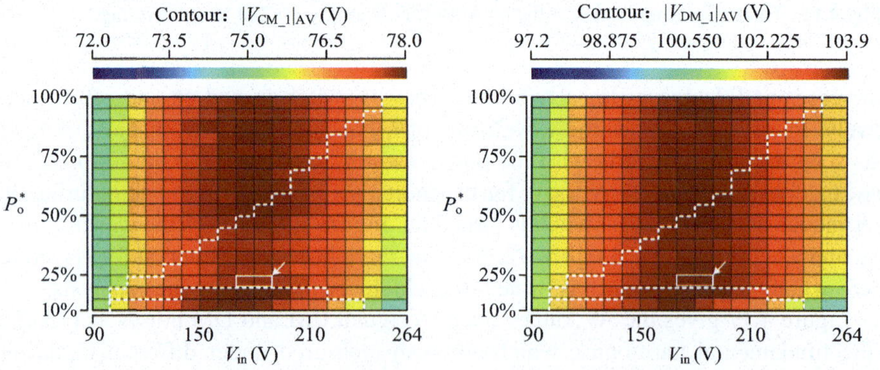

(a) CM noise in AV @175kHz (dBμV) (b) DM noise in AV @175kHz (dBμV)

Fig. 4.17 Measured CM and DM values in AV at the fundamental harmonic

Figure 4.19 shows the measured PK and AV values of the original CM and DM noises at 350 kHz (the second harmonic). It also shows that the conducted EMI of the converter is independent of load and only related with the input voltage in the all-CCM, and the worst case of the harmonic appear under the high input voltage and light load conditions, which are well in agreement with the analysis in Sects. 4.3.2.

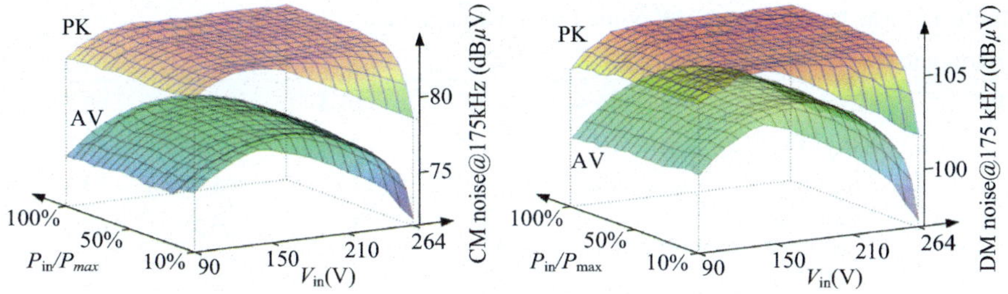

(a) CM noise (b) DM noise

Fig. 4.18 Measured PK and AV values of the fundamental harmonic

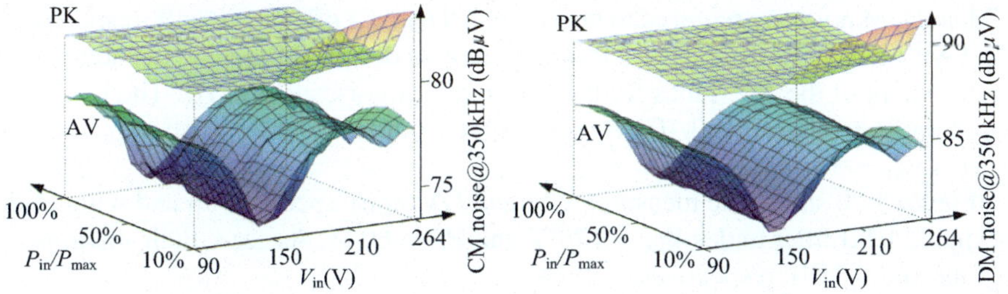

(a) CM noise (b) DM noise

Fig. 4.19 Measured PK and AV values of the second harmonic

Tables 4.2 and 4.3 give the measured largest PK and AV values for the harmonics of CM and DM noise from 150 kHz to 1 MHz and their working conditions. As seen, the difference between the PK and AV values is lower than 10 dB. In addition, the worst case for the CM and DM noise occurs with the input voltage $V_{in} = 170$ V and load $P_o = 60$ W, and the converter is operated in part-CCM/DCM. Therefore, the input voltage condition in the experiment is quite close to satisfy $V_{in}/V_o = 0.45$. More specifically, the maximum AV values for the CM and DM noise at 175 kHz are measured as 78.1 dBμV and 103.9 dBμV, respectively.

Table 4.2 The largest PK and AV of the CM noise harmonics

Harmonic order k		1	2	3	4	5
PK	Max. (dBμV)	83.3	83.3	83.6	83.3	82.0
	V_{in} (V)	160	264	264	264	250
	P_{in} (W)	60	30	30	30	60
AV	Max. (dBμV)	78.1	79	78.9	78.4	78.1
	V_{in} (V)	170	230	240	240	90
	P_{in} (W)	60	30	30	30	75

Table 4.3 The largest PK and AV of the DM noise harmonics

Harmonic order k		1	2	3	4	5
PK	Max. (dBμV)	106.4	91.2	81.3	74.9	69.7
	V_{in} (V)	170	264	264	220	220
	P_{in} (W)	195	30	30	45	45
AV	Max. (dBμV)	103.9	85.9	74.4	66.2	60.8
	V_{in} (V)	170	230	240	240	90
	P_{in} (W)	60	30	30	30	60

In EN55022 standard, the AV limit at 175 kHz is 54.7 dBμV. When calculating the attenuation requirement for the EMI filter, by considering the 6 dB margin, the CM and DM noise are required to be attenuated by 29.4 dB and 55.2 dB, respectively. An EMI filter, as shown on the left in Fig. 4.14, is designed based on the attenuation requirement of the AV values for the CM and DM noise at 175 kHz. The EMI filter topology is given in Fig. 3.10, where, $L_{CM} = 5.6$ mH, $C_y = 2200$ pF, $C_{x1} = C_{x2} = 2\ \mu$F, $L_{DM} = 50\ \mu$H.

Figure 4.20 shows the measured CM and DM noise spectra with and without the designed EMI filter under $V_{in} = 170$ V and $P_o = 60$ W. As seen, their margins are 6.1 dB and 6.3 dB, respectively.

Figure 4.21 shows the measured CM and DM noise spectra with and without the designed EMI filter under $V_{in} = 220$ V and full-load working conditions. As seen, the calculated AV values are in well agreement with the measured values. The margin of the CM and DM noises in Fig. 4.21 is larger than that in Fig. 4.20, which verifies that the CM and DM noise spectra measured under $V_{in} = 170$ V and $P_o = 60$ W are the worst spectra.

Referring to the experimental results, if the EMI filter is designed under the typical 220-V input voltage and full-load condition, the conducted EMI under $V_{in} = 170$ V and $P_o = 60$ W will not satisfy the standard. Therefore, it can be concluded that the CM and DM noise spectra under $V_{in} = 170$ V and $P_o = 60$ W are the worst for average-current-controlled boost PFC converter, which concides with the analysis in Sect. 4.4.3.

4.6 Summary

In this chapter, the input voltage and load condition for the average-current-controlled boost PFC converter operating in all-CCM, part-CCM/DCM, and all-DCM are firstly derived. Based on the STFT and the principle of the EMI receiver, the spectrum of the drain-to-source voltage v_{DS} across power switch is predicted for these operation modes. The worst case for the fundamental harmonic of v_{DS} appears in part-CCM/DCM under $V_{in}/V_o = 0.45$. For the other harmonics, the worst case is more likely to appear in part-CCM/DCM and all-DCM under high input voltage and light-load

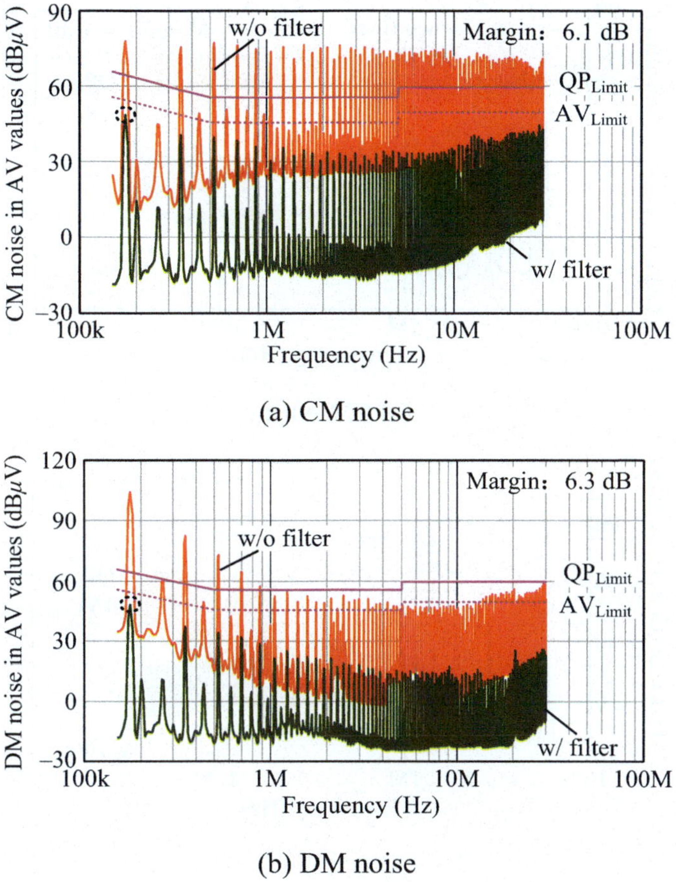

Fig. 4.20 Measured largest AV spectra at $V_{in} = 170$ V and $P_o = 60$ W

conditions. By designing the EMI filter according to the worst case of the first harmonic after 150 kHz, the conducted EMI of average-current-controlled boost PFC converter can satisfy the limit under all working conditions. Furthermore, the approach in this chapter can be also applied for the boost PFC converter operating in DCM.

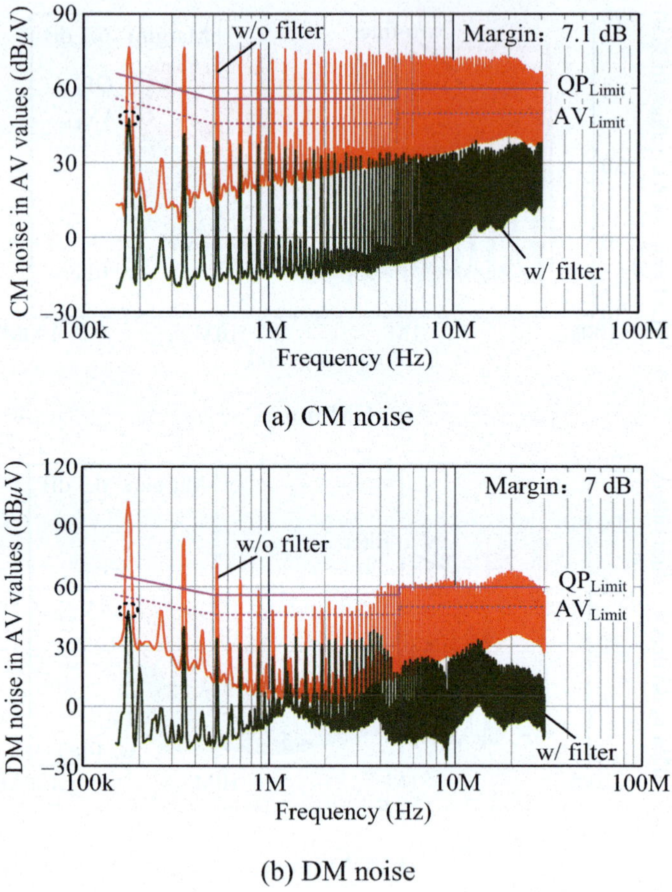

Fig. 4.21 Measured largest AV spectra under full-load with $V_{in} = 220$ V

References

1. R. Tripathi, S. Das, G. Dubey, Mixed-mode operation of boost switch-mode rectifier for wide range of load variations. IEEE Trans. Power Electron. **17**(6), 999–1009 (2002)
2. J.-W. Shin, B.-H. Cho, Digitally implemented average current-mode control in discontinuous conduction mode PFC rectifier. IEEE Trans. Power Electron. **27**(7), 3363–3373 (2012)
3. C. Clark, F. Musavi, W. Eberle, Digital DCM detection and mixed conduction mode control for boost PFC converters. IEEE Trans. Power Electron. **29**(1), 347–355 (2014)
4. V. Lopez, F. Azcondo, A. Castro, R. Zane, Universal digital controller for boost CCM power factor correction stages based on current rebuilding concept. IEEE Trans. Power Electron. **29**(7), 3818–3829 (2014)
5. C. Silva, UC3854 controlled power factor correction circuit design. Unitrode Corp., Application Note U-134 (1999)

6. J. Allen, Short term spectral analysis, synthesis, and modification by discrete Fourier transform. IEEE Trans. Acoust. Speech Signal Process. **25**(3), 235–238 (1977)
7. Z. Wang, S. Wang, P. Kong, F.C. Lee, DM EMI noise prediction for constant on-time, critical mode power factor correction converters. IEEE Trans. Power Electron. **27**(7), 3150–3157 (2012)

Chapter 5
Prediction of the Conducted Electro-magnetic Interference for Critical Conduction Mode Boost Power Factor Correction Converter

5.1 Operation Principle of the CRM Boost PFC Converter

For the boost power factor correction (PFC) converter operated in critical conduction mode (CRM), the switching frequency varies in a half line cycle, and the conducted electromagnetic interference (EMI) spectra are quite complicated, resulting in difficulties for the EMI filter design. In this chapter, the conducted EMI spectra characteristics and EMI filter design for the CRM boost PFC converter will be investigated.

Figure 5.1 shows the main circuit and control diagram for the CRM boost PFC converter, where, $D_1 \sim D_4$ constitute the input bridge rectifier; boost inductor L_b, power switch Q_b, diode D_b, and output capacitor C_o constitute the boost converter; R_{Ld} is the load resistor, C_{xac} and C_{xdc} are the differential-mode (DM) capacitors for eliminating the mixed-mode noise. Similar to the average current control, the CRM boost converter includes the output voltage closed-loop and the inductor current closed-loop. The output voltage V_o is sensed and compared to the reference signal V_{o_ref}, and the error signal is sent into the voltage regulator $G_v(s)$. The output of $G_v(s)$ is multiplied with the sensed rectified input voltage, which serves as the peak current reference signal. When the inductor current touches the peak current reference, Q_b is turned off; and when the inductor current drops to zero, the zero-crossing detector (ZCD) outputs a high level to turn on Q_b.

Figure 5.2 depicts the waveforms of inductor current i_{Lb} and the drain-to-source voltage of Q_b, v_{DS}, in a half line cycle. Obviously, i_{Lb} is a triangular waveform in a switching period, whose average value is half of its peak value. By keeping the peak value of i_{Lb} sinusoidal, the average current follows a sinusoidal pattern naturally, and the PFC is achieved.

© The Author(s), under exclusive license to Springer Nature Singapore Pte Ltd. 2024

X. Ruan et al., *Conducted Electromagnetic Interference in Power Converters: Modeling, Prediction and Reduction*, CPSS Power Electronics Series,
https://doi.org/10.1007/978-981-99-9295-9_5

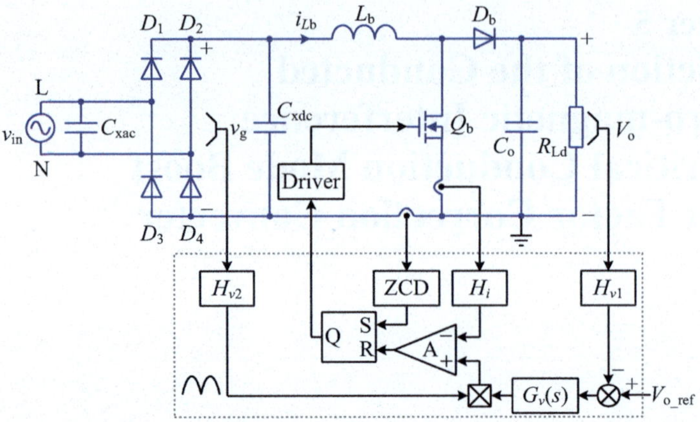

Fig. 5.1 The CRM boost PFC converter

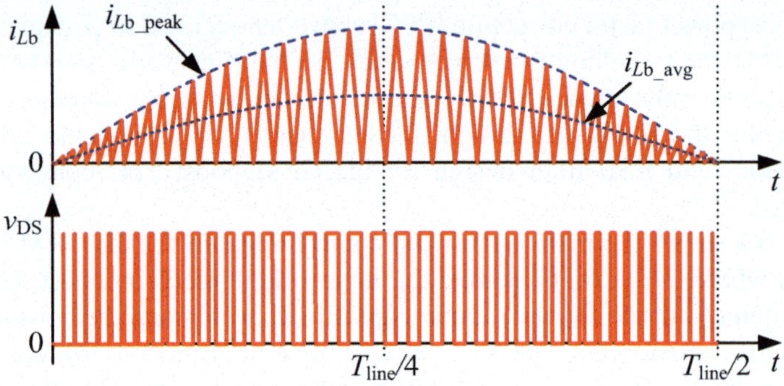

Fig. 5.2 The waveform of inductor current i_{Lb} and the drain-to-source voltage v_{DS}

The rectified input voltage of the boost PFC converter can be expressed as

$$v_g(t) = |v_{in}(t)| = \sqrt{2}V_{in}|\sin \omega_{in}t| \tag{5.1}$$

where, v_{in} is the input voltage, V_{in} is the root-mean-square value of the input voltage, and ω_{in} is the angular frequency of the input voltage.

Under CRM, the duty cycle D_y of the power switch is given by

$$D_y(t) = 1 - \frac{v_g(t)}{V_o} = 1 - \frac{\sqrt{2}V_{in}|\sin \omega_{in}t|}{V_o} \tag{5.2}$$

When Q_b is conducting, v_g is applied on the inductor, forcing i_{Lb} to increase linearly, and the peak value of i_{Lb} is

$$i_{Lb\_peak}(t) = \frac{v_g(t)}{L_b} D_y(t) T_s(t) = \frac{v_g(t) D_y(t)}{L_b f_s(t)} \tag{5.3}$$

where, T_s is the switching period, and f_s is the switching frequency.

According to (5.3), the average inductor current in a switching cycle is

$$i_{Lb\_avg}(t) = \frac{1}{2} i_{Lb\_peak}(t) = \frac{v_g(t) D_y(t)}{2 L_b f_s(t)} \tag{5.4}$$

Since the power factor (PF) equals to 1, $i_{Lb\_avg}(t)$ can be also expressed as

$$i_{Lb\_avg}(t) = \sqrt{2} \frac{P_o}{V_{in}} |\sin \omega_{in} t| = \frac{P_o}{V_{in}^2} v_g(t) \tag{5.5}$$

where, P_o is the output power.

According to (5.1), (5.2), (5.4) and (5.5), the switching frequency f_s for the CRM boost PFC converter is derived as

$$f_s(t) = \frac{1}{2} \frac{V_{in}^2}{P_o L_b} D_y(t) = \frac{1}{2} \frac{V_{in}^2}{P_o L_b} \left(1 - \frac{\sqrt{2} V_{in} |\sin \omega_{in} t|}{V_o} \right) \tag{5.6}$$

According to (5.6), the minimum and maximum switching frequencies in a half line cycle appear when $|\sin\omega_{in} t|$ equals to 1 and 0, respectively, expressed as

$$f_{s\_min} = \frac{1}{2} \frac{V_{in}^2}{P_o L_b} \left(1 - \frac{\sqrt{2} V_{in}}{V_o} \right) \tag{5.7}$$

$$f_{s\_max} = \frac{1}{2} \frac{V_{in}^2}{P_o L_b} \tag{5.8}$$

5.2 Conducted EMI Spectra of the CRM Boost PFC Converter

5.2.1 Harmonic Spectrum of v_{DS}

The drain-to-source voltage of Q_b, v_{DS}, can be treated as the noise voltage source of the conducted EMI in boost PFC converter. As both D_y and f_s change in a half line cycle [1–4], the short-time Fourier transform (STFT) is adopted to analyze the spectrum of v_{DS}.

By applying the STFT to v_{DS} in a switching period, the magnitude for the kth ($k = 1, 2, 3, \ldots$) harmonic with the fundamental frequency of $f_s(t)$, denoted as $V_{DS\_k}(t)$, is expressed as

$$V_{DS\_k}(t) = \left| \frac{1}{T_s} \int_{D_y(t)T_s}^{T_s} V_o e^{-j2k\pi \frac{t_s}{T_s}} dt_s \right| = \frac{2V_o}{k\pi} \left| \sin\left(k\pi D_y(t)\right) \right|, \quad \left(0 \le t \le \frac{T_{line}}{2}\right) \tag{5.9}$$

According to (5.6) and (5.8), we have $D_y(t) = f_s/f_{s\_max}$. By substituting it into (5.9), the harmonic spectrum $V_{DS\_k}(kf_s)$ of v_{DS} at the harmonic frequency of kf_s is derived as

$$V_{DS\_k}(kf_s) = \frac{2V_o}{k\pi} \left| \sin\left(k\pi \frac{f_s}{f_{s\_max}}\right) \right| \tag{5.10}$$

According to (5.9), $V_{DS\_k}(t)$ within $[0, T_{line}/2]$ is shown in Fig. 5.3a. As seen, all these curves are symmetrical to $T_{line}/4$. According to (5.10), $V_{DS\_k}(kf_s)$ is depicted, as shown in Fig. 5.3b. The spectrum characteristics of v_{DS} are summarized below:

(1) For each harmonic, $V_{DS\_k}(kf_s)$ is distributed within $[kf_{s\_min}, kf_{s\_max}]$, and $V_{DS\_k}(kf_{s\_max})$ is always equal to zero.
(2) Within $[f_{s\_min}, f_{s\_max}]$, the amplitude of the fundamental harmonic is larger than that of other harmonics, as highlighted in Fig. 5.3b; Within $[(k-1)f_{s\_max}, kf_{s\_max}]$ ($k \ge 2$), the amplitude of the kth harmonic is the largest.

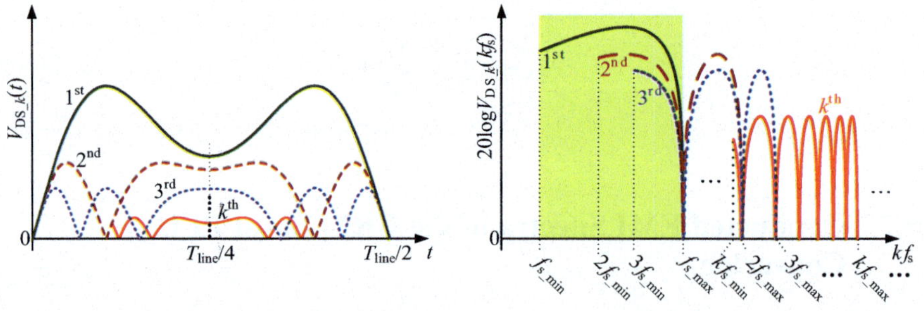

(a) The harmonic amplitude of $V_{DS\_k}(t)$ (b) The spectrum of $V_{DS\_k}(kf_s)$

Fig. 5.3 The harmonic amplitude of $V_{DS\_k}(t)$ and the spectrum of $V_{DS\_k}(kf_s)$

5.2.2 Harmonic Spectra of the CM and DM Noise Voltages

As described in Chap. 3, the harmonic spectra of the common-mode (CM) and DM noise, V_{CM_k} and V_{DM_k}, equal to the harmonic spectrum $V_{DS\_k}(kf_s)$ of v_{DS} multiplied by the CM and DM voltage transfer gain $|CMTG(kf_s)|$ and $|DMTG(kf_s)|$, respectively, i.e.,

$$\begin{cases} V_{CM\_k} = |\mathrm{CMTG}(kf_s)| \cdot V_{DS\_k}(kf_s) = \pi k f_s C_p R_{LN} \cdot V_{DS\_k}(kf_s) \\ V_{DM\_k} = |\mathrm{DMTG}(kf_s)| \cdot V_{DS\_k}(kf_s) = \frac{1}{8\pi^2 (kf_s)^2 L_b C_{x1}} \cdot V_{DS\_k}(kf_s) \end{cases} \quad (5.11)$$

Substituting (5.10) into (5.11), leads to

$$\begin{cases} V_{CM\_k} = 2V_o f_s C_p R_{LN} \left| \sin\left(k\pi \frac{f_s}{f_{s\_max}}\right) \right| \\ V_{DM\_k} = \frac{V_o}{4k^3 \pi^3 f_s^2 L_b C_{x1}} \left| \sin\left(k\pi \frac{f_s}{f_{s\_max}}\right) \right| \end{cases} \quad (5.12)$$

According to (5.12), the harmonic spectra of V_{CM_k} and V_{DM_k} can be depicted in logarithmic scale, as shown in Fig. 5.4. Similar to the spectrum characteristics of v_{DS}, $V_{CM\_k}(kf_{s\_max})$ and $V_{DM\_k}(kf_{s\_max})$ are equal to 0; within $[f_{s\_min}, f_{s\_max}]$, V_{CM_1} and V_{DM_1} are larger than the amplitudes of the other harmonics; and within $[(k-1)f_{s\_max}, kf_{s\_max}]$ $(k \geq 2)$, V_{CM_k} and V_{DM_k} are the largest.

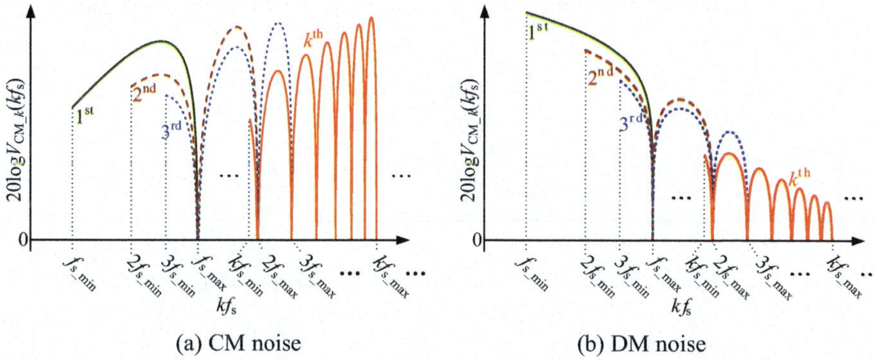

Fig. 5.4 The spectrum of each harmonic for the CM and DM noise

5.2.3 *Peak, Quasi-peak, and Average Spectra of the Conducted EMI*

Based on the principle of EMI receiver in Fig. 2.7, the CM and DM noises of the CRM boost PFC converter are sent to the equivalent intermediate-frequency (IF) filter. As a result, the kth harmonic amplitudes of the CM and DM noises filtered by the equivalent intermediate frequency (IF) filter are

$$\begin{cases} V'_{\mathrm{CM}\_k} = |G_{\mathrm{IF}}(kf_s)| \cdot V_{\mathrm{CM}\_k} \\ V'_{\mathrm{DM}\_k} = |G_{\mathrm{IF}}(kf_s)| \cdot V_{\mathrm{DM}\_k} \end{cases} \tag{5.13}$$

where, G_{IF} is the gain of the IF filter.

The outputs of the envelope detector, $V_{\mathrm{CM\_env}}$ and $V_{\mathrm{DM\_env}}$, are expressed as

$$\begin{cases} V_{\mathrm{CM\_env}} = \max\left\{V'_{\mathrm{CM}\_1}, V'_{\mathrm{CM}\_2}, \ldots, V'_{\mathrm{CM}\_k}\right\} \\ V_{\mathrm{DM\_env}} = \max\left\{V'_{\mathrm{DM}\_1}, V'_{\mathrm{DM}\_2}, \ldots, V'_{\mathrm{DM}\_k}\right\} \end{cases} \tag{5.14}$$

According to (5.6), (5.12) and (5.14), Fig. 5.5 depicts the switching frequency f_s and its multiples, the harmonic spectrum of the CM or DM noise voltages, the envelope signals ($V_{\mathrm{CM\_env}}$ and $V_{\mathrm{DM\_env}}$ in logarithmic scale) at a certain frequency f_{IF} ($f_{\mathrm{IF}} \in [f_{s\_\min}, f_{s\_\max}]$), and the peak (PK) values, where $t_{\mathrm{kfs=IF}}$ denotes the time instant when $kf_s = f_{\mathrm{IF}}$ within [0, $T_{\mathrm{line}}/4$]. As seen, $V_{\mathrm{CM\_env}}$ and $V_{\mathrm{DM\_env}}$ are nearly zero at most of the time within [0, $T_{\mathrm{line}}/2$], except at those time instants $t_{\mathrm{kfs=IF}}$.

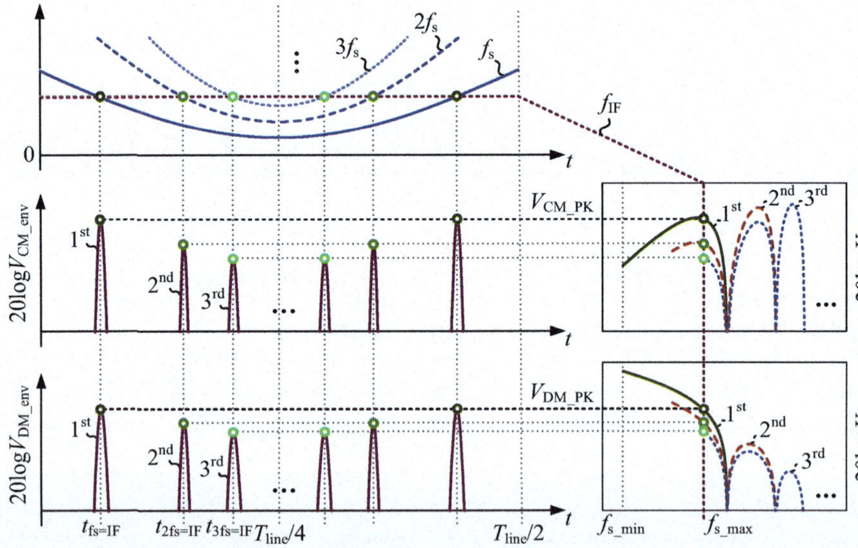

Fig. 5.5 Envelope signal and PK values of the CM and DM noises

Recalling that the gain of the IF filter for the kth harmonic of the CM and DM noises is unity at $t_{kfs=IF}$, i.e., $|G_{IF}[kf_s(t_{kfs=IF})]| = 1$, and other harmonics outside the resolution bandwidth (RBW) of the IF filter are greatly attenuated. Therefore, at $t_{fs=IF}$, according to (5.13) and (5.14), V_{CM_env} and V_{DM_env} are equal to the amplitudes of the fundamental harmonics of the CM and DM noise voltages, respectively, i.e., $V_{CM\_env} = V_{CM\_1}$ and $V_{DM\_env} = V_{DM\_1}$. At $t_{kfs=IF}$, $V_{CM\_env} = V_{CM\_k}$ and $V_{DM\_env} = V_{DM\_k}$, as shown in Fig. 5.5.

1. PK Spectrum

As mentioned in Sect. 5.2.2, within $[f_{s\_min}, f_{s\_max}]$, the fundamental harmonic of the CM and DM noise voltages are larger than other harmonics. Since the PK detector outputs the maximum value of the envelope signal within $[0, T_{line}/2]$, when $f_{IF} \in [f_{s\_min}, f_{s\_max}]$, the PK values of the CM and DM noises, V_{CM_PK} and V_{DM_PK}, are equal to V_{CM_1} and V_{DM_1}, respectively. Likewise, referring to Fig. 5.5, when $f_{IF} \in [(k-1)f_{s\_max}, kf_{s\_max}]$ $(k \geq 2)$, $V_{CM\_PK} = V_{CM\_k}$ and $V_{DM\_PK} = V_{DM\_k}$.

2. QP and AV Spectrum

Figure 5.6 shows the switching frequency f_s, the envelope signal V_{CM_env} (V_{DM_env}) in a half line cycle and the detection of the PK, QP, and average (AV) values, where, T_{BW} is the duration when the switching frequency falls into the 9 kHz RBW of the IF filter. The PK detection is irrelevant to T_{BW}, while the charging process of the QP detector occurs within T_{BW}. With the increase of T_{BW}, the QP value will also be increased to approach the PK value. Therefore, the longer the T_{BW} is, the closer the QP value and the PK value will be. As the RBW of the IF filter is fixed, T_{BW} increases with the decrease of $|df_s(t)/dt|$ at $t_{fs=IF}$. When $f_{IF} = f_{s\_min}$, T_{BW} is the longest over a half line cycle. As a result, the QP value is the closest to the PK value, as shown in Fig. 5.6b.

The AV values of the CM and DM noises are the average value of V_{CM_env} and V_{DM_env} in a half line cycle, respectively, which can be expressed as

$$\begin{cases} V_{CM\_AV} = \frac{2}{T_{line}} \int_0^{\frac{T_{line}}{2}} V_{CM\_env}(t)\,dt \\ V_{DM\_AV} = \frac{2}{T_{line}} \int_0^{\frac{T_{line}}{2}} V_{DM\_env}(t)\,dt \end{cases} \tag{5.15}$$

Figure 5.7 shows the harmonic amplitudes of the CM and DM noise voltages in solid line and the PK, QP, and AV spectra of the CM and DM noises in dashed line, where, the QP values are calculated using the numerical method [5], and the AV values are calculated using (5.15). As seen, the curves of the PK values coincide with the fundamental harmonics within $[f_{s\_min}, f_{s\_max}]$, and they are also identical to the kth harmonics within $[(k-1)f_{s\_max}, kf_{s\_max}]$ $(k \geq 2)$. Note that the harmonic amplitudes are zero at the frequencies lower than f_{s_min}, but the PK, QP, and AV values are not zero (i.e., $-\infty$ dBμV) when f_{IF} is lower than f_{s_min}. This is due to the band-pass effect of the IF filter, whose gain cannot achieve $-\infty$ dB outsides its

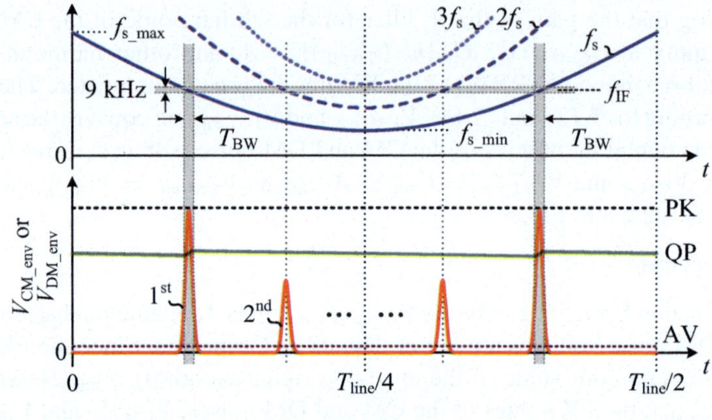

(a) $f_{IF} \neq f_{s\_min}$

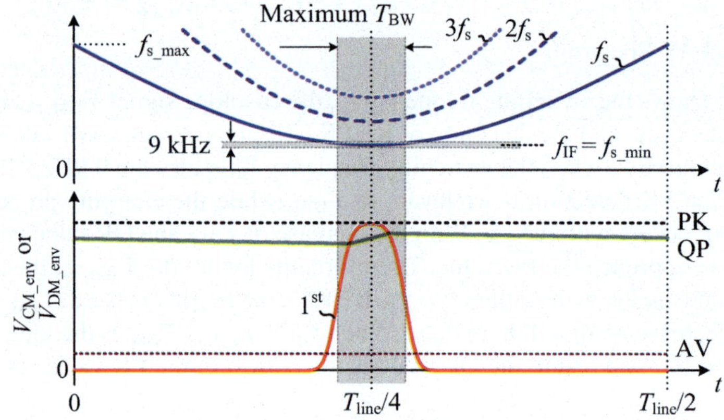

(b) $f_{IF}=f_{s\_min}$

Fig. 5.6 The switching frequency, envelope signal, PK, QP and AV values of the conducted EMI

RBW. Similarly, the PK, QP, and AV values are slightly higher than the amplitudes of the harmonics at the vicinity of kf_{s_max}, owing to the band-pass effect of the IF filter.

Referring to Fig. 5.6, all the harmonics appearing in the envelope signal contribute to the AV values, and with the increase of f_{IF}, the kth harmonic starts to appear in the envelope signal at kf_{s_min}, so there are bulges on the curves of the AV values at the vicinity of kf_{s_min}, as shown in Fig. 5.7. In addition, the switching frequency of the CRM boost PFC converter varies in a wide range, which is far wider than the RBW of the IF filter, so V_{CM_env} and V_{DM_env} are nearly zero at most time within [0, T_{line}/2], and the AV values of the conducted EMI of the CRM boost PFC converter are far lower than the QP and PK values (usually more than 10 dB) [5]. Since the

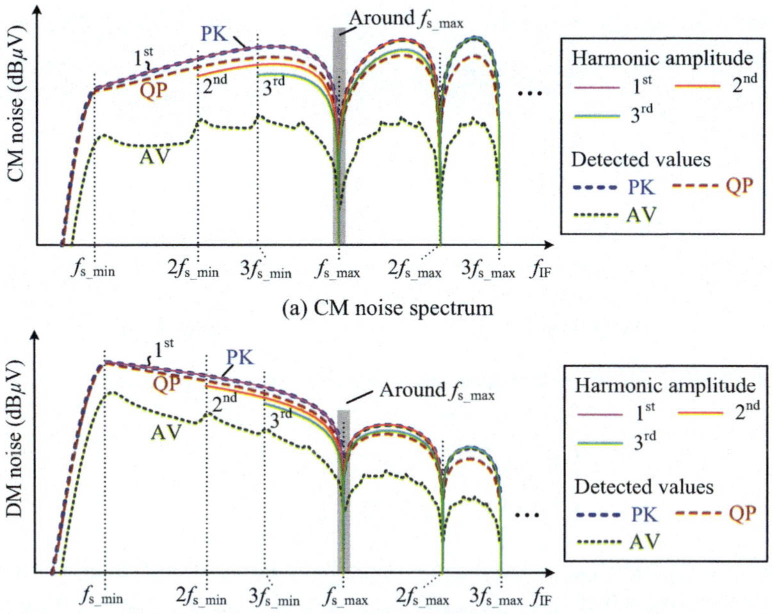

(a) CM noise spectrum

(b) DM noise spectrum

Fig. 5.7 Harmonic amplitudes and the PK, QP, AV values of the conducted EMI

QP limit defined in EN55022 Class B is 10 dB higher than the AV limit, the EMI filters required for suppressing the QP spectra are larger than that for suppressing the AV spectra. So, further discussion will be focused on the worst QP spectra, which require the lowest corner frequencies for the CM and DM filters.

5.3 Prediction of the Conducted EMI of the CRM Boost PFC Converter

As discussed in Sect. 5.2.3, when $f_{IF} \in [f_{s\_min}, f_{s\_max}]$, V_{CM_PK} equals to V_{CM_1} and V_{DM_PK} equals to V_{DM_1}. Substituting $k = 1$ and $f_s = f_{IF}$ into (5.12), the PK values measured within $[f_{s\_min}, f_{s\_max}]$ can be expressed as

$$\begin{cases} V_{CM\_PK} = 2V_o f_{IF} C_p R_{LN} \left|\sin\left(\pi \frac{f_{IF}}{f_{s\_max}}\right)\right| \\ V_{DM\_PK} = \frac{V_o}{4\pi^3 f_{IF}^2 L_b C_{x1}} \left|\sin\left(\pi \frac{f_{IF}}{f_{s\_max}}\right)\right| \end{cases} \quad f_{IF} \in \left[f_{s\_min}, f_{s\_max}\right] \tag{5.16}$$

Since $|\sin(\pi f_{IF}/f_{s\_max})| \leq 1$, according to (5.16), we have

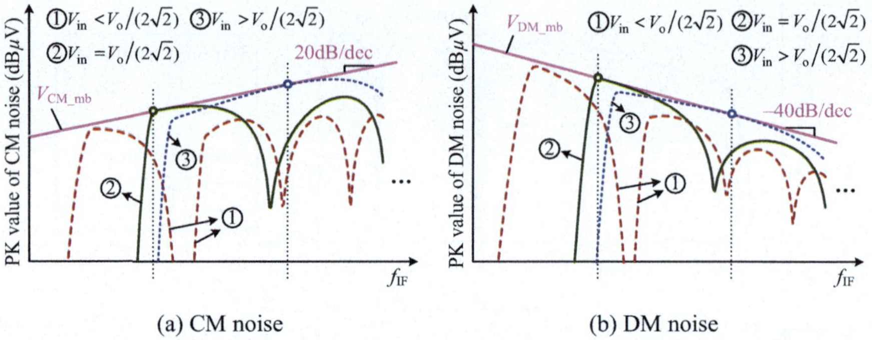

Fig. 5.8 PK value of CM and DM noise spectrum and the maximum boundary

$$\begin{cases} V_{\text{CM\_PK}} \leq 2V_o f_{\text{IF}} C_p R_{\text{LN}} \triangleq V_{\text{CM\_mb}} \\ V_{\text{DM\_PK}} \leq \frac{V_o}{4\pi^3 f_{\text{IF}}^2 L_b C_{x1}} \triangleq V_{\text{DM\_mb}} \end{cases} \tag{5.17}$$

where, V_{CM_mb} and V_{DM_mb} are the maximum boundaries of the CM and DM noises. Based on (5.16), when $f_{IF} = f_{s\_max}/2$, V_{CM_PK} and V_{DM_PK} equals to V_{CM_mb} and V_{DM_mb}, respectively.

According to (5.16), the PK and QP spectra for the CM and DM noise calculated under the same load and different input voltage conditions are depicted in Fig. 5.8.

According to (5.7) and (5.8), it can be known that:

(1) When $V_{in} < V_o/\left(2\sqrt{2}\right)$, we have $f_{s\_max}/2 < f_{s\_min}$, i.e., $f_{s\_max}/2$ does not stay within $[f_{s\_min}, f_{s\_max}]$. Therefore, according to (5.16), the PK values for the CM and DM noises are lower than V_{CM_mb} and V_{DM_mb}, respectively;
(2) When $V_{in} = V_o/\left(2\sqrt{2}\right)$, we have $f_{s\_min} = f_{s\_max}/2$, and the PK values for the CM and DM noise at f_{s_min} are equal to V_{CM_mb} and V_{DM_mb}, respectively;
(3) When $V_{in} \geq V_o/\left(2\sqrt{2}\right)$, the PK values of CM and DM noises are equal to V_{CM_mb} and V_{DM_mb} at $f_{IF} = f_{s\_max}/2$, respectively. Furthermore, $f_{s\_max}/2$ increases with the increase of V_{in} and decrease of P_o.

Therefore, V_{CM_mb} and V_{DM_mb} can be regarded as the maximum boundaries for the PK spectra, as shown in Fig. 5.8. According to (5.17), V_{CM_mb} increases with a slope of 20 dB/dec, and V_{DM_mb} decreases with a slope of – 40 dB/dec. According to (5.8), the frequencies at which the PK values equal to the maximum boundaries can be derived as

$$f_{\text{PK=mb}} = \frac{1}{2} f_{\text{s\_max}} = \frac{1}{4} \frac{V_{\text{in}}^2}{P_o L_b} \quad \left(V_{\text{in}} \geq \frac{V_o}{2\sqrt{2}} \right) \tag{5.18}$$

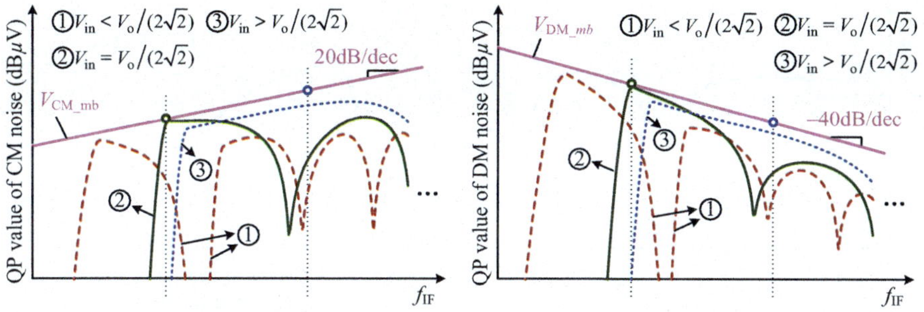

(a) CM noise (b) DM noise

Fig. 5.9 QP value of CM and DM noise spectrum and the maximum boundary

With numerical calculation, the QP values of the CM and DM noises under different input voltage and load conditions are shown in Fig. 5.9. The QP values are calculated by an iterative algorithm to find the equilibrium output when the charging and discharging process of the QP detector is balanced. When $V_{in} = V_o/\left(2\sqrt{2}\right)$, we have $f_{s\_min} = f_{s\_max}/2$. At the minimum switching frequencies, the PK values of the CM and DM noises are equal to V_{CM_mb} and V_{DM_mb}, respectively. As discussed in Sect. 5.2.3, when $f_{IF} = f_{s\_min}$, the QP values are the closest to the PK values. As seen in Fig. 5.9, when $V_{in} < V_o/\left(2\sqrt{2}\right)$ or $V_{in} > V_o/\left(2\sqrt{2}\right)$, the QP values for the CM and DM noises are lower than V_{CM_mb} and V_{DM_mb}, respectively.

Substituting $V_{in} = V_o/\left(2\sqrt{2}\right)$ into (5.18), the frequency at which the QP values are the closest to the maximum boundaries V_{CM_mb} and V_{DM_mb} can be derived as

$$f_{QP=mb} = \frac{1}{4}\frac{V_{in}^2}{P_o L_b} = \frac{1}{32}\frac{V_o^2}{P_o L_b} \tag{5.19}$$

According to (5.19), $f_{QP=mb}$ increases with the decrease of P_o. So, the minimum value of $f_{QP=mb\_min}$ is expressed as

$$f_{QP=mb\_min} = \frac{1}{32}\frac{V_o^2}{P_{o\_max} L_b} \tag{5.20}$$

For the CRM boost PFC converter, the theoretical calculated switching frequency of the CRM boost PFC converter may be very high at extremely light load, and it is usually limited by the control IC. For the frequencies higher than the maximum of $f_{QP=mb}$, the maximum boundaries for the second and third harmonics of the CM and DM noises can be derived in a similar way. Obviously, these boundaries are always lower than the extension line of V_{CM_mb} and V_{DM_mb}, respectively. According to (5.16), the QP spectrum with different output power P_o with $V_{in} = V_o/\left(2\sqrt{2}\right)$ can be illustrated, as shown in Fig. 5.10.

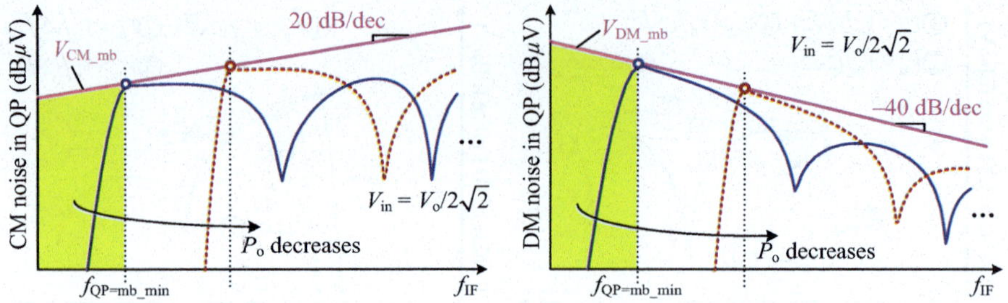

(a) CM noise (b) DM noise

Fig. 5.10 The QP spectrum and the maximum boundary under different load condition

In conclusion, for the frequencies $f_{IF} < f_{QP=mb\_min}$, the QP values measured at f_{IF} are lower than V_{CM_mb} and V_{DM_mb} under all working conditions, as highlighted in the left shadow in Fig. 5.10; and for the frequencies $f_{IF} \geq f_{QP=mb\_min}$, the QP values are the closest to V_{CM_mb} and V_{DM_mb} at $f_{QP=mb}$ when $V_{in} = V_o/\left(2\sqrt{2}\right)$. Thus, V_{CM_mb} and V_{DM_mb} can also be approximately regarded as the maximum boundaries for the QP spectra of the conducted EMI of the CRM boost PFC converter under all working conditions.

5.4 EMI Filter Design for the CRM Boost PFC Converter

When designing the CM and DM filters for CRM boost PFC converters, the attenuation requirements for the CM and DM noises are essential. The maximum boundaries of the QP spectra will be incorporated to derive the attenuation requirement of EMI filters.

Figure 5.11a shows the QP limit in EN55022 Class B [6] and the maximum boundaries of the QP spectra for the CM and DM noises when $f_{QP=mb\_min} \leq 150$ kHz, where, V_{CM_mb} and V_{DM_mb} are the precise maximum boundaries of the CM and DM noises. As seen, the limit decreases in a slope of – 20 dB/dec within [150 kHz, 500 kHz], and V_{CM_mb} increases with a slope of 20 dB/dec and V_{DM_mb} decreases with a slope of – 40 dB/dec, respectively.

By subtracting the QP limit from V_{CM_mb} and V_{DM_mb}, the attenuation requirements for the CM and DM noises, A_{req_CM} and A_{req_DM}, are calculated, respectively, as shown in Fig. 5.11b. It is observed that A_{req_CM} increases with a slope of 40 dB/dec, and A_{req_DM} decreases with a slope of – 20 dB/dec.

With typical CM and DM filters given in Fig. 3.10, the attenuation for the CM and DM noises with a slope of 40 dB/dec can be achieved. According to Fig. 5.11, the attenuation requirement of the CM noise within [150 kHz, 500 kHz] indicates the same lowest corner frequencies of the CM filter, and the lowest corner frequencies of the DM filter are determined by the attenuation requirement of the DM noise at

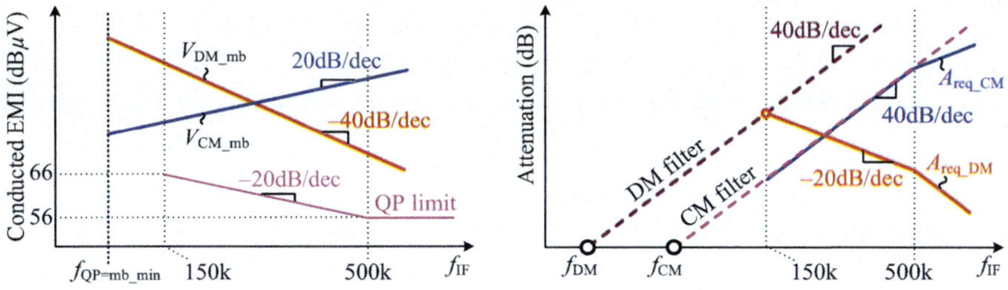

(a) Maximum boundary of the QP spectra and the QP limit (b) Attenuation requirement

Fig. 5.11 CM and DM filter design with $f_{QP=mb\_min} \leq 150$ kHz

150 kHz. Thus, the largest QP values of the CM noises within [150 kHz, 500 kHz] and the largest QP value of the DM noise at 150 kHz indicate the worst CM and DM noise spectra, respectively. According to (5.19), the input voltage and load condition for the worst CM and DM spectra are derived as

$$V_{in\_ws\_}CM = V_{in\_ws\_}DM = \frac{V_o}{2\sqrt{2}} \tag{5.21}$$

$$\frac{V_o^2}{32L_b \cdot 500\,\text{kHz}} \leq P_{o\_ws\_}CM \leq \frac{V_o^2}{32L_b \cdot 150\,\text{kHz}} \tag{5.22}$$

$$P_{o\_ws\_}DM = \frac{V_o^2}{32L_b \cdot 150\,\text{kHz}} \tag{5.23}$$

When 150 kHz $\leq f_{QP=mb\_min} \leq$ 500 kHz, the maximum boundaries of the CM and DM noises and the EMI filter design are shown in Fig. 5.12, where, V'_{CM_mb} and V'_{DM_mb} are the actual boundaries of the CM and DM noises for the frequencies below $f_{QP=mb\_min}$, respectively. Based on (5.16) and (5.17), it is clear that these boundaries are lower than the extension lines of V_{CM_mb} and V_{DM_mb}, respectively.

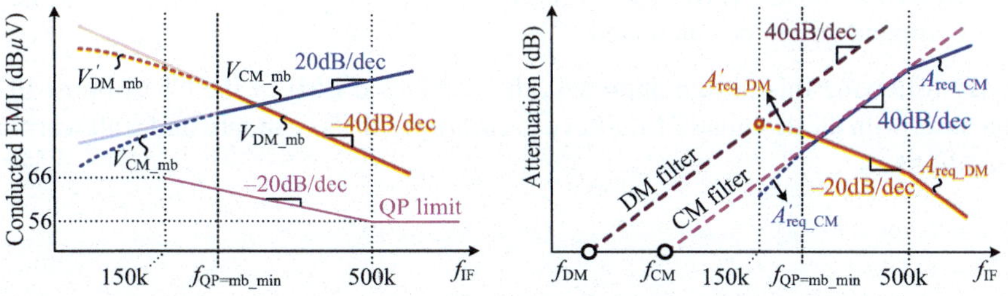

(a) Maximum boundary of the QP spectra and the QP limit (b) Attenuation requirement

Fig. 5.12 CM and DM filter design with 150 kHz $\leq f_{QP=mb\_min} \leq$ 500 kHz

As seen in Fig. 5.12b, the corner frequency of the CM filter is determined by V_{CM_mb} within [$f_{QP=mb\_min}$, 500 kHz], and that of the DM filter is determined by V'_{DM_mb} at 150 kHz. Thus, the input voltage and load conditions for the worst CM noise spectra are: $V_{in\_ws\_CM} = V_o/\left(2\sqrt{2}\right)$ with the output power that ensuring $f_{QP=mb} \leq 500$ kHz. For the DM noise, by substituting $f_{IF} = f_{s\_min} = 150$ kHz and $V_{DM\_QP} = V_{DM\_PK}$ along with (5.7) and (5.8) into (5.12), we have

$$V_{DM\_QP\_150\,kHz} = \frac{2V_o}{\pi}\left|\sin\left[\pi\left(1 - \frac{\sqrt{2}V_{in}}{V_o}\right)\right]\right| \cdot \frac{1}{8\pi^2 L_b C_{x1}(150\text{ kHz})^2} \tag{5.24}$$

where, $V_{in} < V_o/\left(2\sqrt{2}\right)$. With the increase of V_{in}, $1 - \sqrt{2}\,V_{in}/V_o$ gets closer to 0.5. So $V_{DM\_QP\_150kHz}$ increases with the increase of V_{in}.

According to (5.7), when $f_{s\_min} = 150$ kHz and $V_{in} < V_o/\left(2\sqrt{2}\right)$, V_{in} is the largest at full load P_{o_max}. So, the working condition for the largest DM noise at 150 kHz is the output power $P'_{o\_ws\_DM} = P_{o\_max}$ and the input voltage $V'_{in\_ws\_DM}$ satisfying the equation

$$\frac{1}{2}\frac{\left(V'_{in\_ws\_DM}\right)^2}{P_{o\_max}L_b}\left(1 - \frac{\sqrt{2}V'_{in\_ws\_DM}}{V_o}\right) = 150\text{ kHz} \tag{5.25}$$

When $f_{QP=mb\_min} > 500$ kHz, the analysis is similar, and it is not discussed here.

The EMI filter design for CRM boost PFC converters can be summarized as follows.

(1) Measure or predict the worst QP values of the CM and DM noises at f_{req_CM} and f_{req_DM}. When $f_{QP=mb\_min} \leq 150$ kHz, we have $f_{req\_CM} = f_{req\_DM} = 150$ kHz; and when 150 kHz $\leq f_{QP=mb\_min} \leq 500$ kHz, we have $f_{req\_CM} = f_{QP=mb\_min}$ and $f_{req\_DM} = 150$ kHz;
(2) Subtract the QP limit from the QP values of the CM and DM noises, so as to obtain the attenuation requirement for CM and DM noises.
(3) Referring to the EMI filter design in Chap. 3, the parameters of the filter components can be calculated.

By following this design approach, the conducted EMI of the CRM boost PFC converter with the designed EMI filter can comply with the standard under all working conditions.

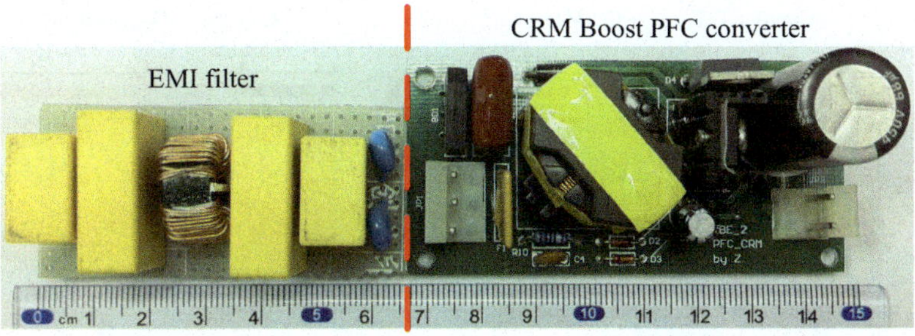

Fig. 5.13 The CRM boost PFC converter prototype

5.5 Experimental Results

5.5.1 Specifications of the Prototype

In order to verify the conducted EMI spectra characteristics of the CRM boost PFC converter, a 90-W CRM boost PFC converter prototype has been fabricated in the lab, as shown in Fig. 5.13.

The specifications of the converter are as follows:

- Input voltage V_{in}: 90~264 V/50 Hz;
- Output voltage V_o: 380 V;
- Rated power P_{o_max}: 90 W;
- Boost inductor L_b: 800 μH (Based on (5.8), the boost inductor is designed to ensure that the minimal switching frequency is above 20 kHz);
- Control IC: L6561 (ST Microelectronics).

5.5.2 Experimental Results

According to (5.20), $f_{QP\_mb\_min}$ is calculated as 62.7 kHz, which is lower than 150 kHz. Therefore, based on (5.21), (5.22) and (5.23), the calculated conditions for the largest QP values of the CM and DM noises at 150 kHz are 134.4 V and 36.1 W, where, the efficiency is 96%. Figure 5.14 show the measured QP values of the CM and DM noises at 150 kHz under different input voltage and load conditions, respectively.

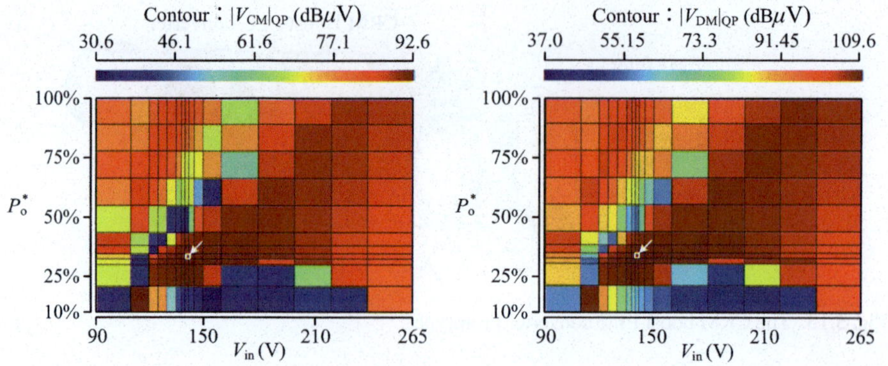

(a) CM noise in QP values at 150 kHz (b) DM noise in QP values at 150 kHz

Fig. 5.14 The measured QP values for the CM and DM noise at 150 kHz

As seen in Fig. 5.14, the measured maximum CM and DM noises are 92.6 and 109.6 dBμV, respectively, both of which occurs at $V_{in} = 140$ V and $P_o = 30$ W, as highlighted in Fig. 5.14. Obviously, the calculated and measured results are quite close, and the difference between the calculated and measured working conditions for the largest QP value at 150 kHz is attributed by the deviation of the duty cycle caused by the voltage drop in the circuit and the delay of the turn-on signal caused by the ZCD of the control IC L6561.

Figure 5.15 gives the measured QP spectra of the CM and DM noises within [150 kHz, 1 MHz], under $V_{in} = 140$ V with $P_o = 15$, 20 and 30 W. As seen, the maximum boundary of the CM noise increases with a slope of 20 dB/dec, and that of the DM noise decreases with a slope of − 40 dB/dec, which agrees well with the analysis in Sect. 5.3.

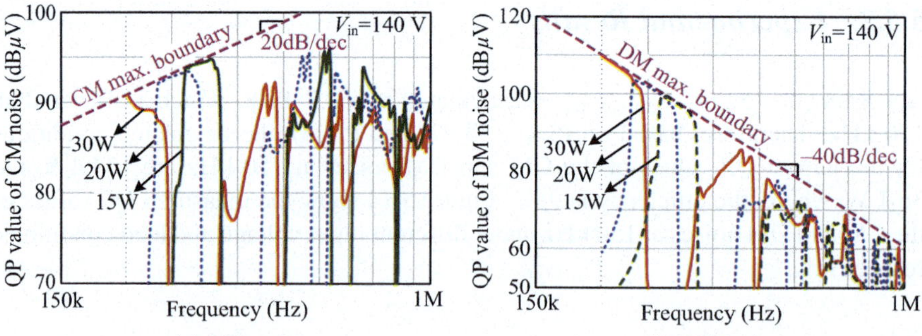

(a) CM noise (b) DM noise

Fig. 5.15 The measured conducted EMI under $V_{in} = 140$ V, $P_o = 15$, 20 and 30 W

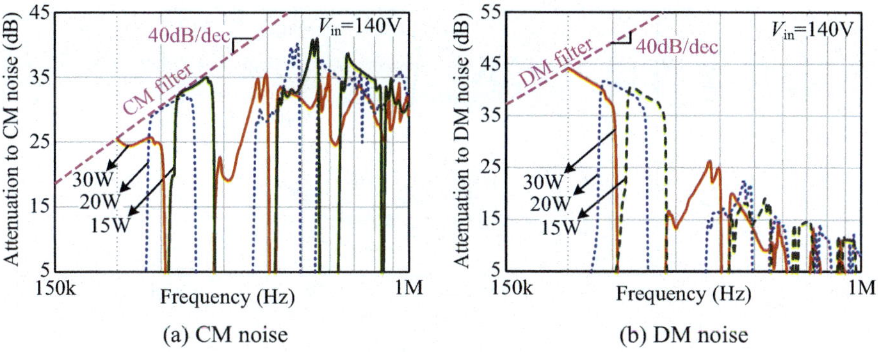

Fig. 5.16 Attenuation requirement for the CM and DM noise

Figure 5.16 gives the calculated attenuation requirements of the spectra in Fig. 5.15. It shows that when the CM and DM filters provide 40-dB/dec attenuation, the attenuation requirement of the CM noise spectra indicate the same corner frequencies of the CM filter, and the corner frequencies of the DM filter are determined by the largest DM noise at 150 kHz, which verifies the theoretical analysis.

As the EN5022 Class B standard is 66 dBμV at 150 kHz, by considering 6 dB margin, 60 dBμV is subtracted from the largest QP values of the CM and DM noises in Fig. 5.14a, b, respectively. As a result, the attenuation requirement of the CM and DM noises at 150 kHz can be calculated as 32.6 and 49.6 dB, respectively. A typical EMI filter topology shown in Fig. 3.10 is adopted to attenuate the conducted EMI, and the components in the filter are designed as follows: $L_{CM} = 11$ mH, $C_y = 2200$ pF, $L_{DM} = 55.7$ μH, $C_{x1} = C_{x2} = 1.47$ μF. The prototype of the EMI filter is given, as shown in the left side of Fig. 5.13.

Figure 5.17 shows the measured CM and DM noise spectra without and with the EMI filter, where the test condition is $V_{in} = 140$ V and $P_o = 30$ W. The measured QP values of the CM and DM noises at 150 kHz are 59.7 and 59.9 dBμV, and the margins of the CM and DM noises are 6.3 and 6.1 dB, respectively.

Figure 5.18 shows the measured CM and DM noise spectra with the EMI filter under $V_{in} = 110$ V and 220 V at full-load, and the margins of the spectra are larger than the spectra of Fig. 5.17. Therefore, it is verified that the EMI spectra of the prototype under $V_{in} = 140$ V and $P_o = 30$ W are worse than that under typical input and full-load conditions.

5.6 Summary

In this chapter, the prediction of the conducted EMI for the CRM boost PFC converter is presented. The harmonic spectra of v_{DS} across the power switch is analyzed, and the characteristics of the CM and DM noises are predicted. It is revealed that the

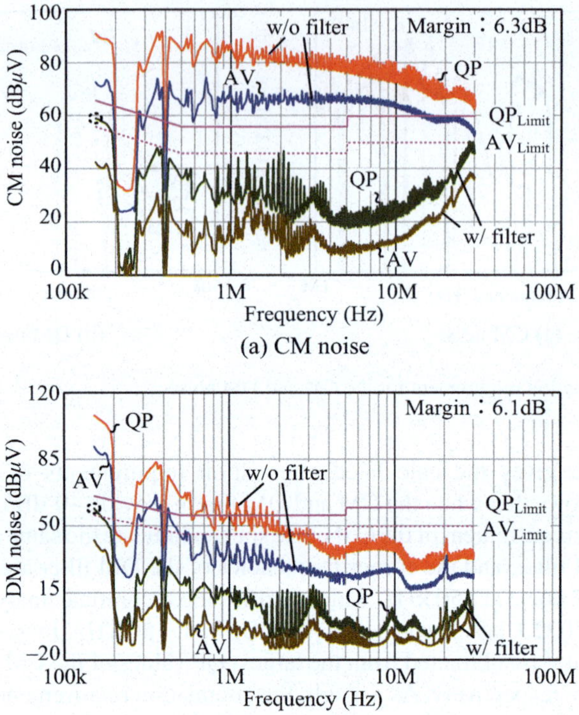

(a) CM noise

(b) DM noise

Fig. 5.17 The measured conducted EMI under $V_{in} = 140$ V and $P_o = 30$ W

EMI filter for the CRM boost PFC converter should be designed to suppress the QP spectra, and the maximum boundaries of the QP spectra under all working conditions are derived. The worst-case conducted EMI spectra of the CRM boost PFC converter is the CM and DM noise spectra with the largest QP values at 150 kHz, when the frequency range of the maximum boundary covers 150 kHz. By designing the EMI filter according to the worst case, the conducted EMI of CRM boost PFC converter can satisfy the limit under all working conditions.

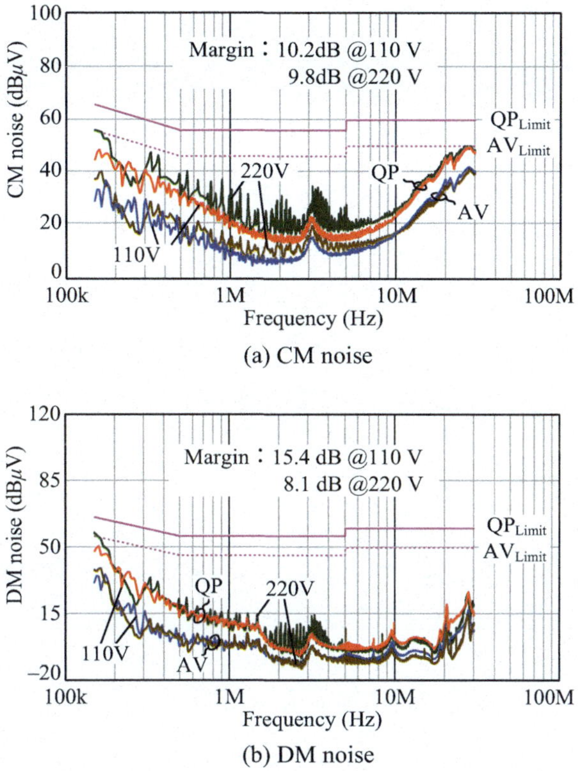

Fig. 5.18 The measured conducted EMI under full-load with $V_{in} = 110$ V and 220 V

References

1. J.-S. Lai, D. Chen, Design consideration for power factor correction boost converter operating at the boundary of continuous conduction mode and discontinuous conduction mode, in *Proceedings of IEEE Applied Power Electronics Conference and Exposition (APEC)* (1993), pp. 267–273
2. J. Zhang, J. Shao, P. Xu, F.C. Lee, M.M. Jovanovic, Evaluation of input current in the critical mode boost PFC converter for distributed power systems, in *Proceedings of IEEE Applied Power Electronics Conference and Exposition (APEC)* (2001), pp. 130–136
3. M. Marvi, A. Fotowat-Ahmady, A fully ZVS critical conduction mode boost PFC. IEEE Trans. Power Electron. **27**(4), 1958–1965 (2012)
4. K. Liu, Y. Lin, Current waveform distortion in power factor correction circuits employing discontinuous-mode boost converters, in *Proceedings of IEEE Power Electronics Specialists Conference (PESC)* (1989), pp. 825–829
5. Z. Wang, S. Wang, P. Kong, F.C. Lee, DM EMI noise prediction for constant on-time, critical mode power factor correction converters. IEEE Trans. Power Electron. **27**(7), 3150–3157 (2012)
6. EN 55022, *Limits and Methods of Measurement of Radio Disturbance Characteristics of Information Technology Equipment*. European Norm Standard, European (2006)

Chapter 6
Modeling of the Common-Mode Noise for Isolated DC-DC Converters

6.1 The CM Noise Propagation Paths of Isolated DC-DC Converters

In order to analyze and reduce the common-mode (CM) noise in isolated DC-DC converters, the CM noise model should be established. Figure 6.1 shows an isolated DC-DC converter, which is composed of the primary circuit, transformer, and the secondary rectifier and filter. The 25-Ω resistor is the equivalent impedance for the CM noise at the line impedance stabilization network (LISN) side, C_{p1} is the parasitic capacitance between the high dv/dt node and the protective earth (PE), C_{ps} is the inter-winding capacitance of the transformer, and PG and SG represents the primary and secondary grounds, respectively.

6.2 A Generalized Lumped Capacitance Model

According to Fig. 6.1, the CM noise is related to the displacement currents flowing through C_{p1} and C_{ps}. Since the inter-winding capacitance C_{ps} is complex, it is usually converted into lumped capacitance.

6.2.1 Characteristics of Inter-winding Capacitance of the Transformer

Figure 6.2 shows a typical transformer winding structure, which are tightly wound. There are layer-to-layer parasitic capacitances between W_{P1} and W_{S1}, W_{S1} and W_{S2}, and W_{S2} and W_{P2}, and the displacement current can flow between these adjacent

© The Author(s), under exclusive license to Springer Nature Singapore Pte Ltd. 2024

X. Ruan et al., *Conducted Electromagnetic Interference in Power Converters: Modeling, Prediction and Reduction*, CPSS Power Electronics Series,
https://doi.org/10.1007/978-981-99-9295-9_6

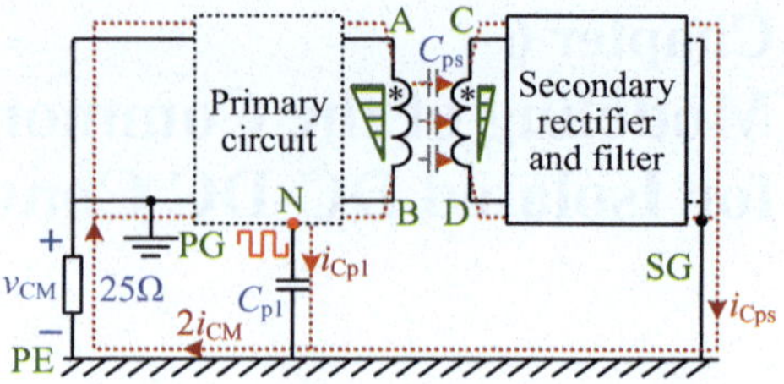

Fig. 6.1 The CM noise propagation path of isolated DC-DC converter

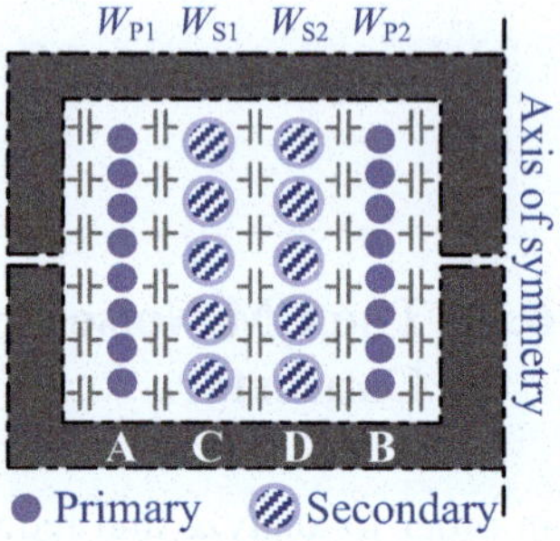

Fig. 6.2 A typical transformer winding structure

layers [1]. Besides, there are also winding-to-core capacitances, including W_{P1}-to-core capacitance and W_{P2}-to-core capacitance. Since the impedance of the core is quite small compared to that of the winding-to-core capacitance, the W_{P1}-to-core and W_{P2}-to-core capacitances are connected in series, and they constitute the layer-to-layer parasitic capacitance between W_{P1} and W_{P2}. Therefore, the outmost layer W_{P1} and innermost layer W_{P2} can be treated as adjacent layers via the core, and there is displacement current between them [2].

Among these displacement currents, only the ones that flow through the primary and secondary winding layers contribute to the CM noise. In Fig. 6.2, the displacement currents flowing through the inter-winding capacitance between W_{P1} and W_{S1} as well as W_{P2} and W_{S2} generate the CM noise, and the displacement currents flowing through the intra-winding capacitance between W_{S1} and W_{S2} as well as W_{P1} and W_{P2} do not cause the CM noise.

6.2.2 *The Displacement Current Flowing Through Inter-winding Capacitances*

When calculating the displacement current flowing through the inter-winding capacitances, the current flowing between every pair of adjacent primary- and secondary-winding layers can be calculated first and then added together. Suppose that the inter-winding capacitance exists between M pairs of layers in the primary and secondary windings, and take the ith pair ($1 \leq i \leq M$) of winding layers W_{Pi} and W_{Si} as an

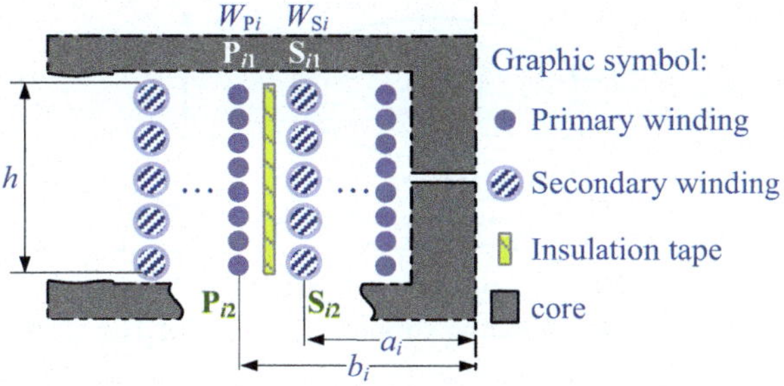

Fig. 6.3 The *i*th pair of the adjacent primary- and secondary-winding layers

example. As shown in Fig. 6.3, the displacement current flowing between W_{Pi} and W_{Si} is related to the inter-winding capacitance and the electric potential distribution along the windings.

The inter-winding capacitance is associated with the dimensions of the winding layers and the medium between the layers. In Fig. 6.3, h is the height of the winding, b_i is the distance between W_{Pi} and the axis of symmetry, and a_i is the distance between W_{Si} and the axis of symmetry. In this configuration, a cylindrical capacitor between W_{Pi} and W_{Si} is formed, which is commonly named as the structural capacitance between layers. When $b_i - a_i \ll h$, the structural capacitance C_{0i} between these two layers is approximately given by [3]

$$C_{0i} = \frac{2\pi\varepsilon_0\varepsilon_r h}{\ln(b_i/a_i)} \tag{6.1}$$

where, ε_0 is the permittivity in the vacuum, and ε_r is the relative permittivity of the dielectric between the windings.

Figure 6.4a shows the primary-winding layer W_{Pi}, where, P_{i1} is the origin terminal, P_{i2} is the end terminal, E is a point in the winding. By neglecting the leakage magnetic flux and AC resistance caused by the skin effect and proximity effect, the electric potential varies linearly along the windings, as shown in Fig. 6.4b, where, v_{Pi1} and v_{Pi2} are the electric potential of P_{i1} and P_{i2} with respect to the PE, respectively, l_{WPi} is the winding length from P_{i1} to P_{i2}, and l_E is the length from E to P_{i2} along the winding.

According to Fig. 6.4b, the electric potential of point E can be expressed as

$$v_E = \frac{v_{Pi1} - v_{Pi2}}{l_{WPi}} l_E + v_{Pi2} \tag{6.2}$$

For the secondary-winding layer W_{Si}, S_{i1} is the origin terminal, S_{i2} is the end terminal, and F is a point in the winding. The length from F to S_{i2} along the winding is denoted by l_F, and the whole winding length from S_{i1} to S_{i2} is denoted by l_{WSi}. Likewise, the electric potential of point F is given by

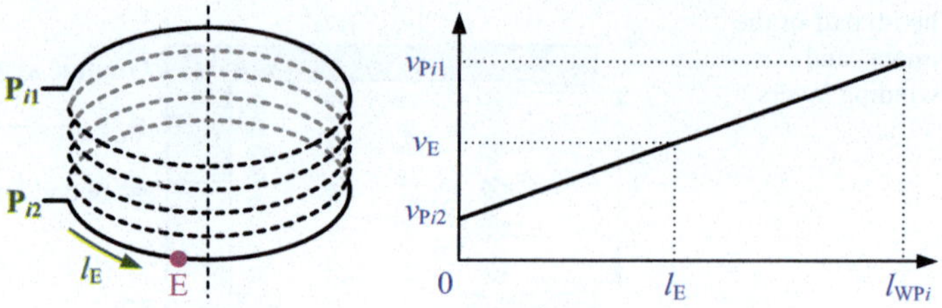

(a) Winding structure (b) The electric potential along the windings.

Fig. 6.4 The structure and electric potential distribution of the transformer winding

$$v_F = \frac{v_{Si1} - v_{Si2}}{l_{WSi}} l_F + v_{Si2} \tag{6.3}$$

The displacement current i_{dis_i} that flows through W_{Pi} and W_{Si} can be calculated with the superposition theory [4]. Firstly, the voltage of W_{Si} is set to zero, and the displacement current i_{pri_i} generated by W_{Pi} can be calculated. Secondly, the voltage of W_{Pi} is set to zero, and the displacement current i_{sec_i} caused by W_{Si} can be obtained. Finally, the total displacement current i_{dis_i} can be obtained by adding i_{pri_i} and i_{sec_i} together. Suppose that the inter-winding capacitance distributes evenly along the transformer windings, the capacitance between W_{Pi} with a unit length and W_{Si} equals to C_{0i}/l_{WPi}, and the capacitance between W_{Si} with a unit length and W_{Pi} equals to C_{0i}/l_{WSi}. Therefore, i_{dis_i} can be derived as

$$i_{dis\_i} = i_{pri\_i} + i_{sec\_i} = \int_0^{l_{WPi}} \frac{C_{0i}}{l_{WPi}} \frac{d}{dt}(v_E - 0) dl_E + \int_0^{l_{WSi}} \frac{C_{0i}}{l_{WSi}} \frac{d}{dt}(0 - v_F) dl_F \tag{6.4}$$

Substitution of (6.2) and (6.3) into (6.4), yields

$$i_{dis\_i} = C_{0i} \frac{d}{dt}\left(\frac{v_{Pi1} + v_{Pi2}}{2} - \frac{v_{Si1} + v_{Si2}}{2}\right) = C_{0i} \frac{d}{dt}(\overline{v}_{WPi} - \overline{v}_{WSi}) \tag{6.5}$$

where, $\overline{v}_{WPi} = (v_{Pi1} + v_{Pi2})/2$, $\overline{v}_{WSi} = (v_{Si1} + v_{Si2})/2$, and they represent the averaged electric potential (AEP) of winding layers W_{Pi} and W_{Si}, respectively.

Based on (6.5), the total displacement current i_{dis} flowing through M pairs of adjacent primary- and secondary-winding layers are obtained as

$$i_{dis} = \sum_{i=1}^{M} i_{dis\_i} = \sum_{i=1}^{M} C_{0i} \frac{d}{dt}(\overline{v}_{WPi} - \overline{v}_{WSi}) \tag{6.6}$$

For the primary winding, A and B denote the origin- and end-terminals, respectively, and the turns from points P_{i1}, P_{i2} to terminal B are represented by N_{Pi1} and N_{Pi2}.

Likewise, for the secondary winding, C and D denote the origin- and end-terminals, respectively, and the turns from points S_{i1}, S_{i2} to terminal D are represented by N_{Si1} and N_{Si2}. According to linear variation of the electric potential along the length of windings, the voltage drop of each turn of the windings is the same. Hence, we have

$$\begin{cases} v_{Pi1} = \frac{v_A - v_B}{N_P} N_{Pi1} + v_B, v_{Pi2} = \frac{v_A - v_B}{N_P} N_{Pi2} + v_B \\ v_{Si1} = \frac{v_C - v_D}{N_S} N_{Si1} + v_D, v_{Si2} = \frac{v_C - v_D}{N_S} N_{Si2} + v_D \end{cases} \tag{6.7}$$

where, N_P and N_S are the total turns number of the primary and secondary windings, respectively, and v_B and v_D are the electric potential of the terminals B and D, respectively.

According to (6.7), we have

$$\begin{cases} \overline{v}_{WPi} = \left(v_{Pi\_1} + v_{Pi\_2}\right)/2 = \lambda_{Pi} v_A + (1 - \lambda_{Pi}) v_B \\ \overline{v}_{WSi} = \left(v_{Si\_1} + v_{Si\_2}\right)/2 = \lambda_{Si} v_C + (1 - \lambda_{Si}) v_D \end{cases} \tag{6.8}$$

where, the coefficient λ_{Pi} and λ_{Si} are expressed as

$$\begin{cases} \lambda_{Pi} = \frac{N_{Pi1} + N_{Pi2}}{2N_P} \\ \lambda_{Si} = \frac{N_{Si1} + N_{Si2}}{2N_S} \end{cases} \tag{6.9}$$

According to (6.9), it is obvious that $0 < \lambda_{Pi} < 1$ and $0 < \lambda_{Si} < 1$.

Substituting (6.8) into (6.6), yields

$$i_{dis} = C_0 \frac{d}{dt} [\lambda_P v_A + (1 - \lambda_P) v_B] - C_0 \frac{d}{dt} [\lambda_S v_C + (1 - \lambda_S) v_D] \tag{6.10}$$

where, C_0, λ_A and λ_C are given by

$$\begin{cases} C_0 = \sum_{i=1}^{M} C_{0i} \\ \lambda_P = \frac{1}{C_0} \sum_{i=1}^{M} \lambda_{Pi} C_{0i} \\ \lambda_S = \frac{1}{C_0} \sum_{i=1}^{M} \lambda_{Si} C_{0i} \end{cases} \tag{6.11}$$

According to (6.11), C_0 is the structural capacitance of the transformer, which equals to the sum of C_{0i}; λ_P and λ_S are defined as the winding structural parameters, which are related to the winding structure. According to (6.9) and (6.11), λ_P and λ_S are in the range of (0, 1).

6.2.3 The General Lumped Capacitance Model of the Transformer

Owing to the complexity of inter-winding capacitance, the lumped capacitance model, as shown in Fig. 6.5, is usually adopted for circuit analysis, which includes six lumped capacitances between every two terminals in the transformer. Since the displacement currents flowing through C_{AB} and C_{CD} do not generate CM noise, C_{AB} and C_{CD} are neglected in the following analysis. In this section, the relationship between the parameters in (6.11) for the inter-winding capacitance and the four lumped capacitance will be established.

In Fig. 6.5, the total displacement current that flows through C_{AC}, C_{AD}, C_{BC} and C_{BD} can be derived as

$$
\begin{aligned}
i_{dis} &= C_{AC}\frac{d}{dt}(v_A - v_C) + C_{AD}\frac{d}{dt}(v_A - v_D) + C_{BC}\frac{d}{dt}(v_B - v_C) + C_{BD}\frac{d}{dt}(v_B - v_D) \\
&= (C_{AC} + C_{AD})\frac{dv_A}{dt} + (C_{BC} + C_{BD})\frac{dv_B}{dt} - (C_{AC} + C_{BC})\frac{dv_C}{dt} - (C_{AD} + C_{BD})\frac{dv_D}{dt}
\end{aligned} \tag{6.12}
$$

where, v_A, v_B, v_C and v_D are the electric potentials of the transformer winding terminals with respect to the PE.

Based on the displacement current consistency, a set of linear algebraic equations can be obtained by equalizing (6.10) and (6.12), i.e.,

$$
\underbrace{\begin{pmatrix} 1\,1\,0\,0 \\ 0\,0\,1\,1 \\ 1\,0\,1\,0 \\ 0\,1\,0\,1 \end{pmatrix}}_{\boldsymbol{A}_N}
\begin{pmatrix} C_{AC} \\ C_{AD} \\ C_{BC} \\ C_{BD} \end{pmatrix}
=
\underbrace{\begin{pmatrix} \lambda_P C_0 \\ (1-\lambda_P)C_0 \\ \lambda_S C_0 \\ (1-\lambda_S)C_0 \end{pmatrix}}_{\boldsymbol{D}(C_0)} \tag{6.13}
$$

where, $\boldsymbol{A}_N$ is the coefficient matrix, $\boldsymbol{D}(C_0)$ is the vector of the given equations.

According to (6.13), the rank of $\boldsymbol{A}_N$ and the rank of the augmented matrix $\boldsymbol{A}_N \vdots \boldsymbol{D}(C_0)$ are both equal to 3, and they are less than 4, which is the number of unknown quantities. Therefore, the solution of (6.13) for the lumped capacitance is

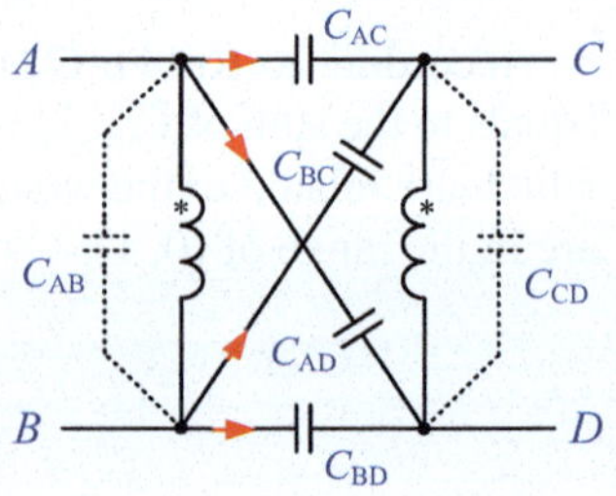

Fig. 6.5 Lumped capacitance of the transformer

not unique, and are given by

$$\begin{pmatrix} C_{AC} \\ C_{AD} \\ C_{BC} \\ C_{BD} \end{pmatrix} = \begin{pmatrix} C \\ \lambda_P C_O - C \\ \lambda_S C_O - C \\ (1 - \lambda_P - \lambda_S) C_O + C \end{pmatrix} \tag{6.14}$$

where, C is a free variable.

6.3 The CM Noise Model of Basic Isolated DC-DC Converters

With the lumped capacitance model of the transformer, the CM noise model of basic isolated dc-dc converters, including flyback, forward, push–pull, half-bridge and full-bridge converters will be derived and discussed in this section.

6.3.1 Derivation for the CM Noise Model of Flyback Converter

Figure 6.6a, b show the flyback converters with output rectifier diode D_o at the high side and low side, respectively. The parasitic capacitance C_{pB} is between the drain of the main switch Q and the PE, and L_m is the magnetizing inductance. i_p is the primary current, i_s is the secondary current, i_{Lm} is the magnetizing current, and i_p equals to $ni_s + i_{Lm}$, where, n is the secondary-to-primary winding turns ratio.

The substitution theorem is adopted to derive the CM noise model by replacing the switches and diodes with voltage source or current source. Note that, in the

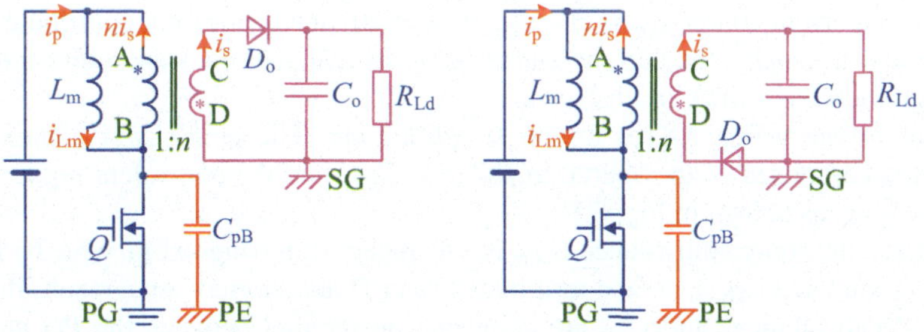

(a) D_o is at the high side (b) D_o is at the low side

Fig. 6.6 The flyback converter with different configurations

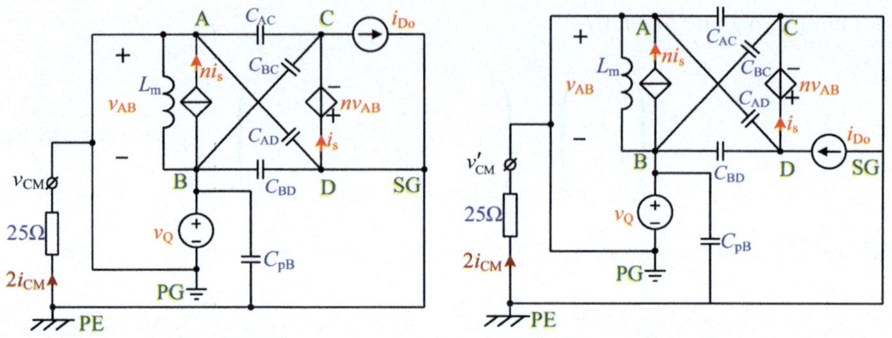

(a) D_o is at the high side (b) D_o is at the low side

Fig. 6.7 The CM noise models of flyback converter

substitution, voltage loops or current nodes in the circuit should be avoided. In addition, the lumped capacitance model is adopted to represent the transformer inter-winding capacitance.

Firstly, switch Q is substituted with voltage source v_Q, which is the AC component of the drain-to-source voltage of switch Q since the DC component does not contribute to the CM noise. Then, the transformer windings are represented by a pair of controlled sources. To avoid voltage loops, the rectifier diode D_o is substituted with current source i_{Do}, which is the AC component of the current through D_o. As the dv/dt of the output voltage is far lower than that of v_Q, the output filter capacitor and load, which are connected in parallel, is treated as short-circuit. Finally, by replacing the inter-winding capacitance with its equivalent lumped capacitance, and considering the 25-Ω equivalent impedance for the CM noise at the LISN side, the CM noise model can be derived, as shown in Fig. 6.7.

The CM noise model in Fig. 6.7 can be further simplified. Taking the CM noise model shown in Fig. 6.7a as an example, superposition theorem is adopted to investigate the effect of voltage and current sources individually. Firstly, to investigate the impact of current source, the voltage source is shorted, thus $v_{AB} = 0$. Accordingly, the current source i_{Do} is essentially shorted, and it does not contribute to the CM noise. Secondly, the current source i_{Do} is opened, and the subcircuit excited by voltage source is shown in Fig. 6.8a.

The circuit in Fig. 6.8a is further simplified into Thevenin's equivalent circuit, which is composed of the source impedance C_{sum} and the equivalent noise source (ENS) v_{ENS}, as shown in Fig. 6.8b.

Firstly, the source impedance C_{sum} is calculated. Referring to Fig. 6.8a, by letting $v_Q = 0$, we have $v_{AB} = 0$, and terminals C and D are shorted. As a result, the four lumped capacitances and C_{pB} are connected in parallel between the PG and SG. According to (6.14), the sum of four lumped capacitance is C_0, so C_{sum} equals to $C_0 + C_{pB}$.

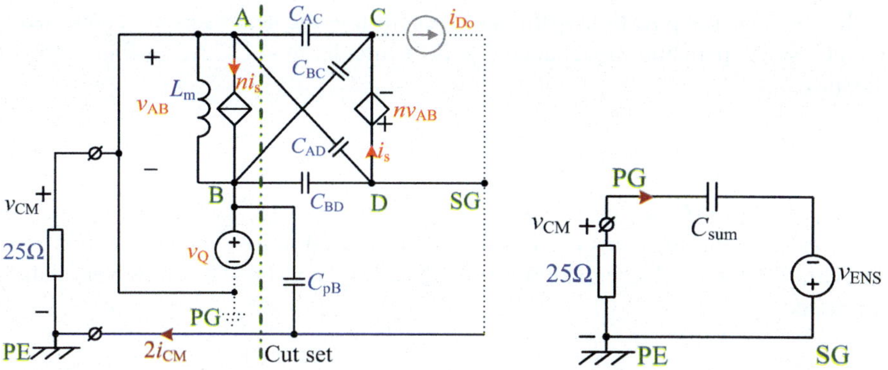

(a) The subcircuit excited by voltage source (b) The simplified model

Fig. 6.8 The simplification of the CM noise model of flyback converter

Secondly, the equivalent noise source v_{ENS}, which is the open-circuit voltage for the network at the right side of the 25-Ω resistor, is calculated. Referring to Fig. 6.8a, by opening the network at the right-hand side of the 25-Ω resistor, the current $2i_{CM}$ flowing from the PE is 0. In Fig. 6.8a, a cut set is drawn from the lumped capacitance to the PE. According to the continuity of current, we have

$$C_{AC}\frac{d}{dt}(v_A - v_C) + C_{AD}\frac{d}{dt}(v_A - v_D) + C_{BC}\frac{d}{dt}(v_B - v_C) + (C_{BD} + C_{pB})\frac{d}{dt}(v_B - v_D) = 0 \tag{6.15}$$

where, v_A, v_B, v_C and v_D is the electric potential of the transformer terminals, respectively.

According to Fig. 6.8a, we have

$$\begin{cases} v_B = v_Q + v_A \\ v_C = v_D - nv_{AB} = v_D + nv_Q \end{cases} \tag{6.16}$$

By substituting (6.16) into (6.15), v_{ENS} is derived as

$$v_{ENS} = v_D - v_A = \frac{(C_{BC} + C_{BD} + C_{pB}) - n(C_{AC} + C_{BC})}{C_{AC} + C_{AD} + C_{BC} + C_{BD} + C_{pB}} v_Q \tag{6.17}$$

Substituting (6.14) into (6.17), leads to

$$v_{ENS} = \frac{(1 - \lambda_P - n\lambda_S)C_0 + C_{pB}}{C_0 + C_{pB}} v_Q \tag{6.18}$$

In low power applications with plastic enclosure, the heat sinks are connected to the PG, so the parasitic capacitance C_{pB} is far smaller than C_0. According to (6.18), we have

$$v_{ENS} \approx (1 - \lambda_P - n\lambda_S)v_Q \tag{6.19}$$

According to (6.19), if $(1 - \lambda_P - n\lambda_S) = 0$, then $v_{ENS} = 0$, implying the CM noise is zero, and the converter has the ability of CM noise cancellation (CMNC). In Sect. 6.4, the CMNC capability of the basic isolated DC-DC converters will be discussed.

6.3.2 The CM Noise Model of Other Basic Isolated DC-DC Converters

Following the derivation of the CM noise model of flyback converter, this section gives the derivation of the CM noise model of other basic isolated DC-DC converters.

1. Forward Converter

Figure 6.9a, b show the forward converters with the output rectifier diode D_{R1} and filter inductor L_f at the high side and low side, respectively. The current flowing through the primary and secondary winding is equal to i_p and i_s, respectively. By neglecting the magnetizing current, i_p equals to ni_s, where n is the secondary-to-primary winding turns ratio.

To derive the CM noise models, switch Q is substituted with voltage source v_Q, which is the AC component of the drain-to-source voltage of switch Q, and the transformer windings are represented by a pair of controlled sources. Then, in order to avoid voltage loops, the reset circuit is replaced by current source i_{RESET}, the rectifier diode D_{R1} is substituted with current source i_{DR1}, and the branch with filter inductor, filter capacitor and load is replaced by ac current source i_{Lf}. These current sources are the AC component of the current through the corresponding branches,

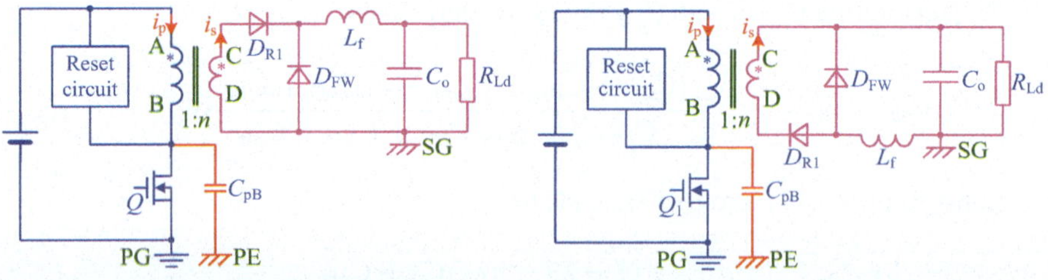

(a) D_{R1} and L_f are at the high side (b) D_{R1} and L_f are at the low side

Fig. 6.9 The forward converter with different configurations

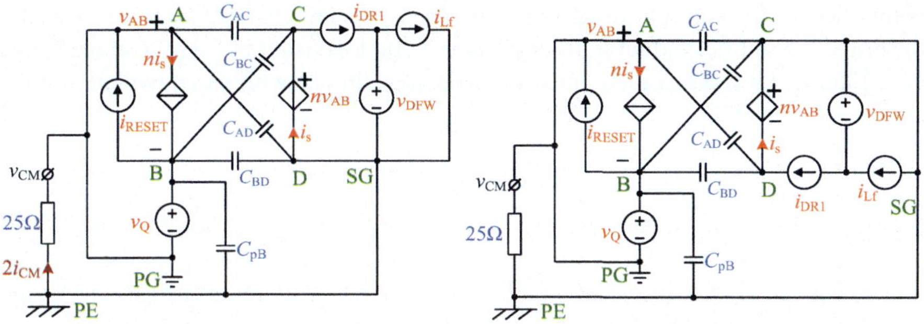

(a) D_{R1} and L_f are at the high side (b) D_{R1} and L_f are at the low side

Fig. 6.10 The CM noise models of forward converters

respectively. To avoid current nodes, the freewheeling diode D_{FW} is substituted with voltage source v_{DFW}, which is the AC component of the voltage across D_{FW}. Finally, the inter-winding capacitance is replaced with its equivalent lumped capacitance, the CM noise models are derived, as shown in Fig. 6.10.

2. Push–Pull Converter

Figure 6.11 shows the push–pull converter with different configurations of the secondary rectifier and filter, where, the parasitic capacitances C_{pA} and C_{pB} are between the drain of the main switches Q_1, Q_2 and the PE, respectively. The currents flowing through the primary and secondary windings are i_{p1}, i_{p2}, i_{s1} and i_{s2}, respectively. By neglecting the magnetizing current, $i_{p1} = ni_{s1}$, and $i_{p2} = ni_{s2}$, where, n is the secondary-to-primary winding turns ratio.

To derive the CM noise models, switches Q_1 and Q_2 are substituted with voltage sources v_{Q1} and v_{Q2}, which are the AC component of the drain-to source voltage of switches Q_1 and Q_2, respectively. The transformer windings are represented by a pair of controlled sources. Then, in order to avoid voltage loops, the rectifier diodes D_{R1} and D_{R2} are substituted with current sources i_{DR1} and i_{DR2}, which are the AC

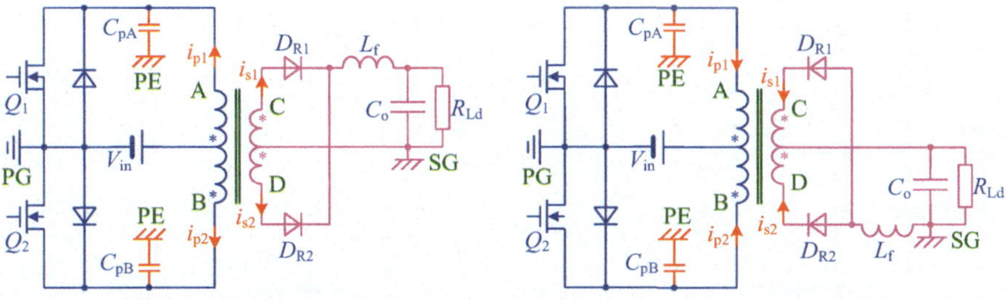

(a) Configuration I (b) Configuration II

Fig. 6.11 The push–pull converter with different configurations component of the current through D_{R1} and D_{R2}, respectively. To avoid current nodes, the branch with filter inductor, filter capacitor and load is replaced by voltage source v_{rect}. Finally, by replacing the inter-winding capacitance with its equivalent lumped capacitance, the CM noise models are derived, as shown in Fig. 6.12.

3. Half-Bridge Converter

Figure 6.13 shows the half-bridge converter with different configurations of the secondary rectifier and filter, where, the parasitic capacitance C_{pB} is between the drain of the main switch Q_2 and the PE. i_p is the primary winding current, and i_{s1} and i_{s2} are the two secondary windings currents. By neglecting the magnetizing current, $i_p = n(i_{s1} - i_{s2})$.

To derive the CM noise models, the switch Q_2 is substituted with voltage source v_{Q2}, which is the AC component of the drain-to-source voltage of Q_2. The transformer windings are represented by a pair of controlled sources. Then, in order to avoid voltage loops, the switch Q_1 is substituted with current source i_{Q1}, and the rectifier diodes D_{R1} and D_{R2} are substituted with current sources i_{DR1} and i_{DR2}. These current

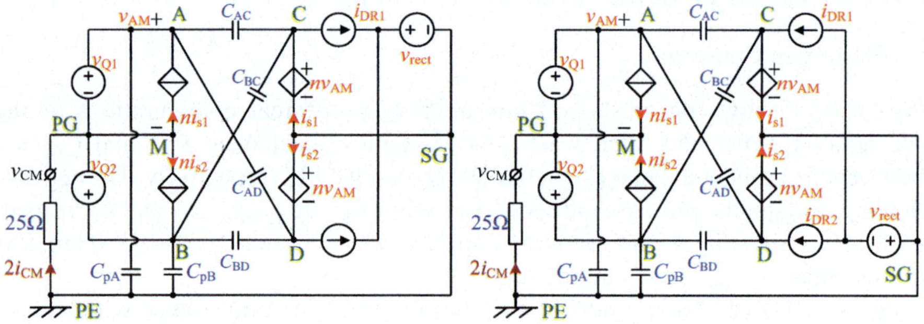

(a) Configuration I (b) Configuration II

Fig. 6.12 The CM noise models of push–pull converters

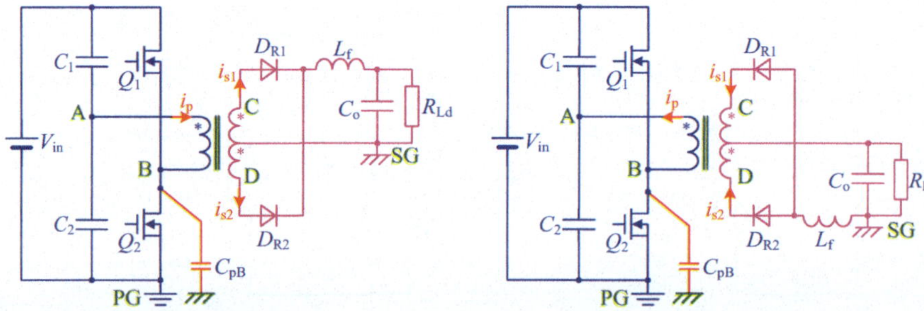

(a) Configuration I (b) Configuration II

Fig. 6.13 The half-bridge converter with different configurations sources are the AC component of the current through the corresponding branches, respectively. To avoid current nodes, the branch with filter inductor, filter capacitor and load is replaced by voltage source v_{rect}. Since the dv/dt of the voltage across C_1 and C_2 is far lower than that of v_{Q2}, these two capacitors can be treated as short circuit. Finally, by replacing the inter-winding capacitance with its equivalent lumped capacitance, the CM noise models are derived, as shown in Fig. 6.14.

4. Full-Bridge Converter

Figure 6.15 shows the full-bridge converter with different configurations of the secondary rectifier and filter, where, the parasitic capacitances C_{pA} and C_{pB} are between the drain of the main switches Q_3, Q_4 and the PE, respectively. i_p is the primary winding current, i_{s1} and i_{s2} are the two secondary windings currents, and $i_p = n(i_{s1} - i_{s2})$.

To derive the CM noise models, the switches Q_3 and Q_4 are substituteded with voltage sources v_{Q3} and v_{Q4}, which are the AC component of the drain-to-source voltage of Q_3 and Q_4, respectively. The transformer windings are represented by a pair of controlled sources. Then, in order to avoid voltage loops, the switches Q_1 and Q_2 are replaced by current source i_{Q1} and i_{Q2}, and the rectifier diodes D_{R1} and

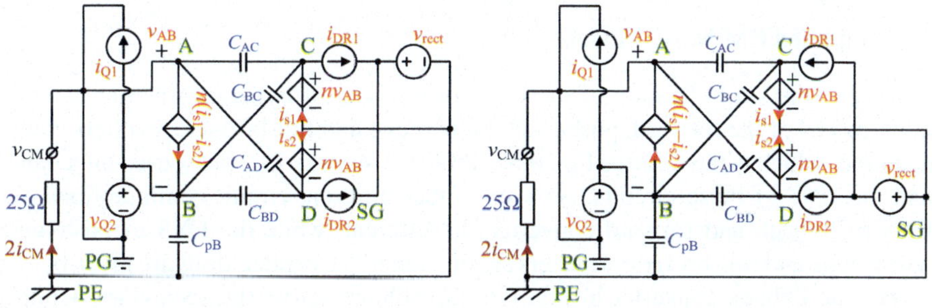

(a) Configuration I (b) Configuration II

Fig. 6.14 The CM noise models of half-bridge converters

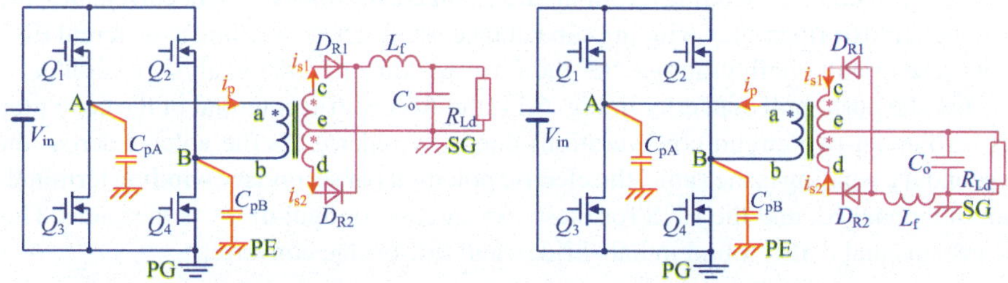

(a) Configuration I (b) Configuration II

Fig. 6.15 The full-bridge converter with different configurations

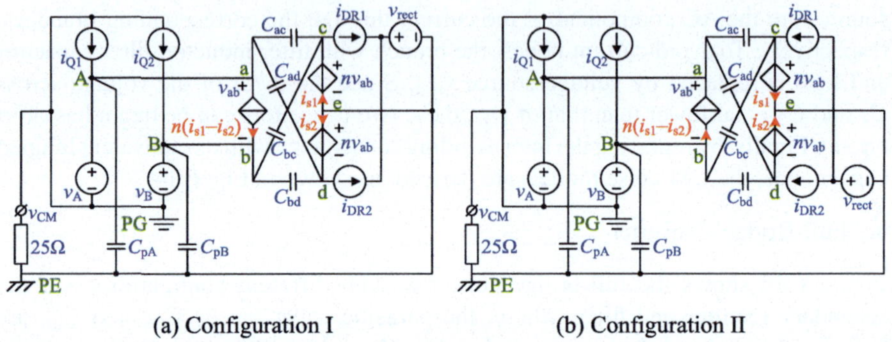

Fig. 6.16 The CM noise models of full-bridge converters

D_{R2} are substituted with current sources i_{DR1} and i_{DR2}. These current sources are the AC component of the current through the corresponding branches, respectively. To avoid current nodes, the branch with filter inductor, filter capacitor and load is replaced by voltage source v_{rect}. Finally, by replacing the inter-winding capacitance with its equivalent lumped capacitance, the CM noise models are derived, as shown in Fig. 6.16.

5. Simplified CM Noise Models

Similar to the simplification of the CM noise model of flyback converter, the CM noise model of the forward, push–pull, half-bridge and full-bridge converters can be acquired, which is the same as Fig. 6.8b. Table 6.1 summarizes the source impedance and the ENS of these circuits. As seen, under different circuit configurations, the ENS of flyback and forward converters is different, while the ENS of push–pull, half-bridge and full-bridge converters are the same. The explanations are given below.

For the flyback converter in Fig. 6.6, the voltage across the secondary winding equals to v_{CD} regardless of the position of D_o. Take the PE as the reference ground, referring to Fig. 6.6a, when D_o is at the high side, the electric potential of terminals C and D is v_{CD} and 0, respectively. Referring to Fig. 6.6b, when D_o is at the low side, the electric potential of terminals C and D is shifted into 0 and $-v_{CD}$, respectively. As the electric potentials of winding terminals are changed, the displacement current flowing through transformer inter-winding capacitance is influenced, leading to different ENS for these circuit configurations. For the forward converter, the analysis is similar.

For the push–pull converter in Fig. 6.11, the AC electric potential of the center tap is zero under two circuit configurations due to symmetry. As the voltage across the secondary winding is the same, the electric potentials of secondary winding terminals are not changed, and the ENS for these two circuit configurations is the same. This scenario can be also found in half-bridge and full-bridge converter.

Table 6.1 The source impedance and ENS of basic isolated DC-DC converters

Topology	Configurations	Source impedance	The ENS formulas
Flyback	D_o is at high side (Fig. 6.6a)	$C_0 + C_{pB}$	$v_{ENS\_FLY1} = \frac{(1-\lambda_P-n\lambda_S)C_0+C_{pB}}{C_0+C_{pB}} v_Q$
	D_o is at low side (Fig. 6.6b)		$v_{ENS\_FLY2} = \frac{[1-\lambda_P+n(1-\lambda_S)]C_0+C_{pB}}{C_0+C_{pB}} v_Q$
Forward	D_{R1} and L_f are at high side (Fig. 6.9a)	$C_0 + C_{pB}$	$v_{ENS\_FWD1} = \frac{(1-\lambda_P+n\lambda_S)C_0+C_{pB}}{C_0+C_{pB}} v_Q$
	D_{R1} and L_f are at low side (Fig. 6.9b)		$v_{ENS\_FWD2} = \frac{[1-\lambda_P-n(1-\lambda_S)]C_0+C_{pB}}{C_0+C_{pB}} v_Q$
Push–pull	Configuration I (Fig. 6.11a)	$C_0 + C_{pA} + C_{pB}$	$v_{ENS\_PP} = \frac{[2\lambda_P-1+n(2\lambda_S-1)]C_0+C_{pA}-C_{pB}}{C_0+C_{pA}+C_{pB}} v_{Q1}$
	Configuration II (Fig. 6.11b)		
Half-bridge	Configuration I (Fig. 6.13a)	$C_0 + C_{pB}$	$v_{ENS\_HB} = \frac{[1-\lambda_P+n(2\lambda_S-1)]C_0+C_{pB}}{C_0+C_{pB}} v_{Q2}$
	Configuration II (Fig. 6.13b)		
Full-bridge	Configuration I (Fig. 6.15a)	$C_0 + C_{pA} + C_{pB}$	$v_{ENS\_FB} = \frac{[\lambda_P + n(1-2\lambda_S)]C_0 + C_{pA}}{C_0 + C_{pA} + C_{pB}} v_{Q3} + \frac{\{1-[\lambda_P + n(1-2\lambda_S)]\}C_0 + C_{pB}}{C_0 + C_{pA} + C_{pB}} v_{Q4}$
	Configuration II (Fig. 6.15b)		

6.4 The CMNC Capability of Isolated DC-DC Converters

If the CM noise in a converter can be cancelled by optimizing the transformer winding structure, it means this converter has the ability of CMNC. In Sect. 6.3.1, the flyback with D_o at the high side is proved to have the CMNC capability. In this Section, the CMNC capability of the basic isolated dc-dc converters will be investigated.

For the flyback converter with D_o at the high side, according to Table 6.1, the coefficient before v_Q in the ENS is $[(1-\lambda_P-n\lambda_S)C_0 + C_{pB}]/(C_0 + C_{pB})$. Since λ_P and λ_S are in the range of (0, 1), this coefficient is between $(-nC_0 + C_{pB})/(C_0 + C_{pB})$ and 1. Similarly, the range of the coefficients for the flyback converter with D_o at the low side, forward and half-bridge converters can be calculated, and they are listed in Table 6.2. For the push–pull converter, Q_1 and Q_2 are usually mounted on the same heat sink, and the insulation in the middle adopts the same material with equal thickness, so C_{pA} and C_{pB} are almost equal, and the range for the coefficient in the ENS is calculated. Likewise, for the full-bridge converter, C_{pA} and C_{pB} can also be regarded as equal. When the pulse width modulation (PWM) control is adopted, the diagonal switches are turned on/off simultaneously, so $v_{Q3} = -v_{Q4}$. Accordingly, the range of the coefficient in the ENS can be calculated, and is also listed in Table 6.2.

According to Table 6.2, for the flyback converter with D_o at the low side, the coefficient before v_Q in the ENS is always positive, and the CM noise is impossible to be cancelled. Similarily, the forward converter with D_{R1} and L_f at the high side has no CMNC capability.

For the flyback converter with D_o at the high side, the coefficient before v_Q in the ENS is between $(-nC_O + C_{pB})/(C_O + C_{pB})$ and 1. Normally, C_O is far larger than C_{pB}, so the term $(-nC_O + C_{pB})/(C_O + C_{pB})$ is a negative number. As the coefficient before v_Q is a continuous function regarding λ_P and λ_S, there exists a solution with λ_P and λ_S that makes the coefficient be zero, indicating that the flyback converter with D_o at the high side has the CMNC capability. Similarily, the forward converter with D_{R1} and L_f at the low side, push–pull, half-bridge and full-bridge converters also have the CMNC capability.

When the full-bridge converter adopts phase-shift control, a resonant inductor L_r is usually added to assist soft-switching. With phase-shift control, v_{Q3} and v_{Q4} are no longer complementary, and the CM noise can not be cancelled by each other. Moreover, the CM noise is affected by the voltage across L_r. In Chap. 8, the reduction of the CM noise in the phase-shifted full-bridge (PSFB) converter will be investigated.

Table 6.2 The range of the ENS of basic isolated dc-dc converters

Topology	Configurations	The range of the coefficient before the voltage source in the ENS	CMNC capability
Flyback	D_o is at high side (Fig. 6.7a)	$\left(\frac{-nC_O+C_{pB}}{C_O+C_{pB}}, 1\right)$	Yes
	D_o is at low side (Fig. 6.7b)	$\left(\frac{C_{pB}}{C_O+C_{pB}}, 1+\frac{nC_O}{C_O+C_{pB}}\right)$	No
Forward	D_{R1} and L_f are at high side (Fig. 6.9a)	$\left(\frac{C_{pB}}{C_O+C_{pB}}, 1+\frac{nC_O}{C_O+C_{pB}}\right)$	No
	D_{R1} and L_f are at low side (Fig. 6.9b)	$\left(\frac{-nC_O+C_{pB}}{C_O+C_{pB}}, 1\right)$	Yes
Push–pull	Configurations in Fig. 6.11	$\left(-\frac{(1+n)C_O}{C_O+C_{pA}+C_{pB}}, \frac{(1+n)C_O}{C_O+C_{pA}+C_{pB}}\right)$	Yes
Half-bridge	Configurations in Fig. 6.13	$\left(\frac{-nC_O+C_{pB}}{C_O+C_{pB}}, 1+\frac{nC_O}{C_O+C_{pB}}\right)$	Yes
Full-bridge	Configurations in Fig. 6.15	$\left(-\frac{2nC_O}{C_O+C_{pA}+C_{pB}}, \frac{(1+2n)C_O}{C_O+C_{pA}+C_{pB}}\right)$	Yes

6.5 Experimental Verification and Discussions

In this section, the lumped capacitance model and the CM noise model will be verified by the experimental results, respectively.

6.5.1 *The Verification of the Lumped Capacitance Model*

Figure 6.17a, b shows the test setup for verifying the lumped capacitance model of the transformer and the corresponding test circuit [5]. The excitation port of the network analyzer provides an excitation source v_{ac} with an internal 50-Ω resistor, which is applied on the primary winding of the transformer. The secondary winding of the transformer is left to be open. A 10-kΩ resistor R is connected across the primary-winding terminal B and secondary-winding terminal D for measuring the displacement current flowing through the transformer inter-winding capacitance. The voltage across R, v_{DB}, is measured by the probe and sent into the response port of the network analyzer. Note that the input impedance of the probe is 10-MΩ resistance and 8-pF capacitance in parallel.

As seen from Fig. 6.17b, due to the continuity of current, the displacement current i_{dis} flowing through the inter-winding capacitance is equal to the current i flowing through the resistor R and the input impedance of the probe, which are connected in parallel. Thus, the generated voltage v_{DB} forms an indirect observation for i_{dis}. The network analyzer adopts Agilent E5061B, which measures the magnitude and phase of the transfer function from v_{AB} to v_{DB} within [150 kHz, 30 MHz] conducted EMI frequency range.

Figure 6.17c shows the transformer under test. The core uses RM 10. The primary winding adopts single strand enameled wire with 0.3 mm, and its turns number $N_P = 29$; The secondary winding uses single strand enameled wire with 0.8 mm, and its turns number $N_S = 11$. The structural capacitance C_0 is measured by LCR meter at 100 kHz. In the measurement, the primary winding and secondary winding are shorted, respectively, and the capacitance between the shorted two windings is C_0. The measured C_0 at 100 kHz is 38.23 pF. Owing to the single-layer structure, the winding structural parameter $\lambda_P = \lambda_S = 0.5$. By substituting them into (6.14), the equivalent lumped capacitances are expressed as

$$C_{AC} = C_{BD} = C,\quad C_{AD} = C_{BC} = \frac{C_0}{2} - C \tag{6.20}$$

Assuming the leakage inductance of the transformer is so small that can be neglected, Fig. 6.18 shows the equivalent circuit of that shown in Fig. 6.17b, where the transformer secondary winding is substituted with a controlled voltage source

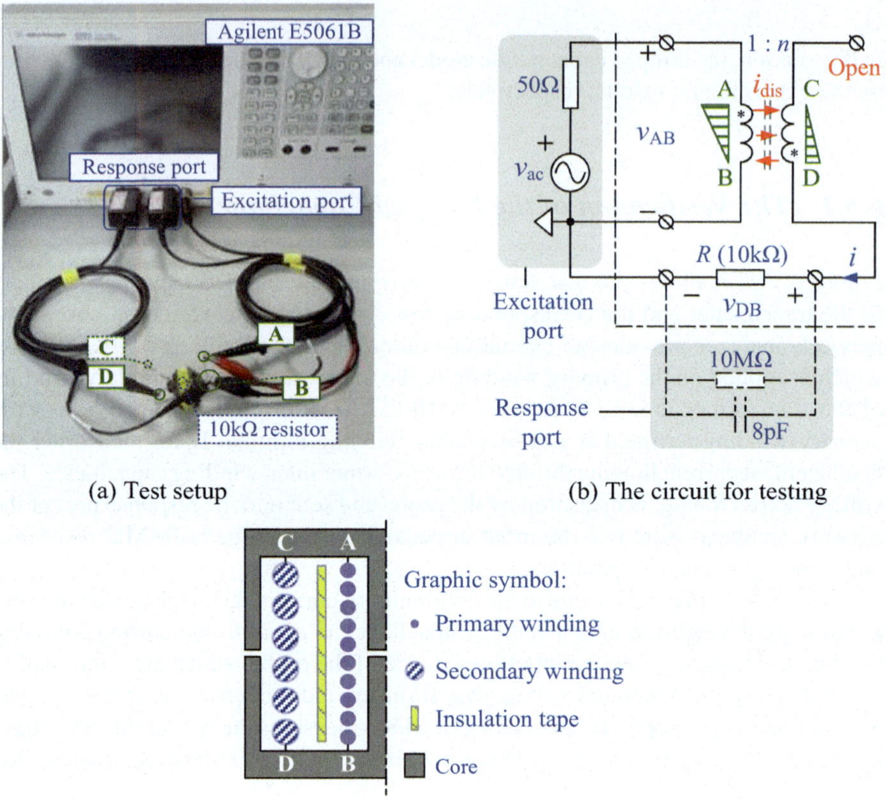

(a) Test setup

(b) The circuit for testing

(c) The winding structure of the transformers under test

Fig. 6.17 The principle of measuring the lumped capacitor of the transformer

nv_{AB}, and $n = N_S/N_P$ is the secondary-to-primary winding turns ratio. For simplifying the calculation, the C in (6.20) is set to $C_0/2$. Thus, we have $C_{AC} = C_{BD} = C_0/2$, and $C_{AD} = C_{BC} = 0$, which are accordingly not shown in Fig. 6.18. In addition, since the 10-MΩ resistance in the input impedance of the probe is far larger than the 10-kΩ resistance, it can be neglected.

According to Fig. 6.18, the transfer function from v_{AB} to v_{DB} can be derived as

$$G_{TR}(s) = \frac{v_{DB}(s)}{v_{AB}(s)} = \frac{1+n}{2}\frac{sRC_0}{sR(C_0 + C_{probe}) + 1} \tag{6.21}$$

Figure 6.19 shows the measured and theoretical curves of $G_{TR}(s)$. As seen, the theoretical and measured curves match well from 150 kHz to 30 MHz, verifying the effectiveness of the lumped capacitance model for the transformer.

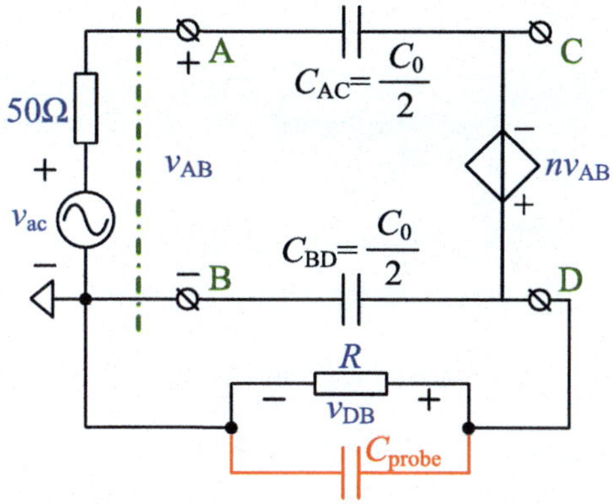

Fig. 6.18 The theoretical model for calculating the transfer function $G(s)$

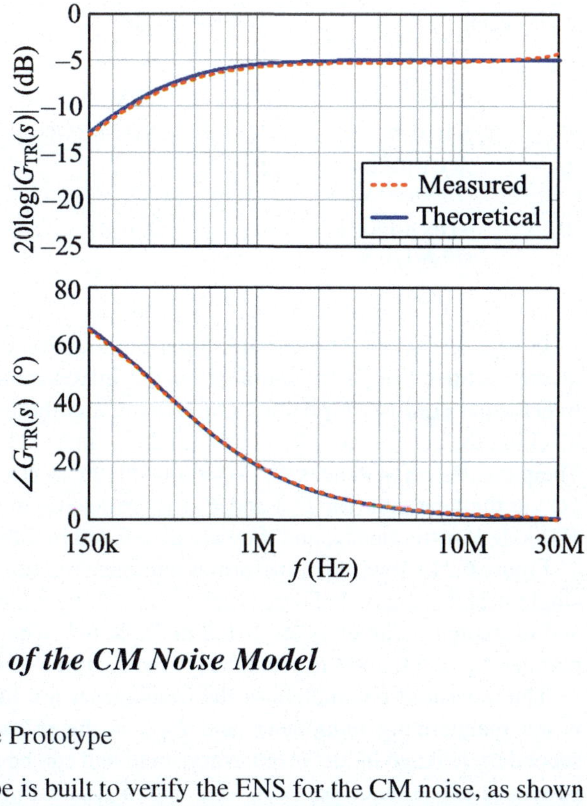

Fig. 6.19 The measured and theoretical transfer function $G_{TR}(s)$ for the transformer

6.5.2 The Verification of the CM Noise Model

1. The Specifications of the Prototype

A flyback converter prototype is built to verify the ENS for the CM noise, as shown in Fig. 6.20a. Figure 6.20b illustrates the the main circuit. The specifications of the prototype are: input voltage v_{ac} = 90~264 V, output voltage V_o = 19.5 V, output power P_o = 40 W. The switching frequency is 66.5 kHz, and the components used in the prototype are listed in Table 6.3.

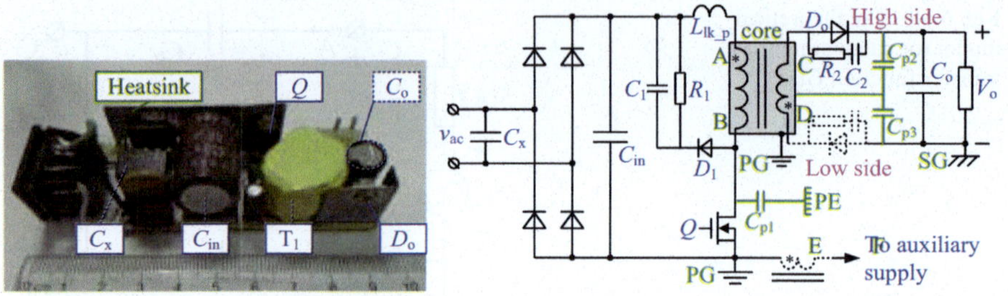

(a) Photo of the prototype (b) The main circuit of flyback converter

Fig. 6.20 The photo and the main circuit of flyback converter

Table 6.3 The key parameters of the prototype and the measured parasitic capacitance

Main circuit		Snubber circuit		Parasitic capacitance	
C_x	0.22 μF	R_1	94 kΩ	C_{p1}	11.00 pF
C_{in}	68 μF	C_1	2200 pF	C_{p2}	5.5 pF
C_o	680 μF	D_1	S1PM-E3/84A	C_{p3}	14.67 pF
Q	AOTF10N60 (10 A/600 V)	R_2	47 Ω		
D_o	STPS20120CT (20 A/120 V)	C_2	220 pF		

In Fig. 6.20b, a 0.22-μF differential-mode (DM) capacitor C_x is added in front of the rectifier bridge for eliminating the mixed-mode noise [6]. In addition, the transformer core is wrapped by copper foil, and the foil is connected to the PG for blocking the electric coupling between the primary winding and the PE via the core. The parasitic capacitances C_{p1} (drain-to-PE), C_{p2} (from the core to the cathode of D_o via the heat sink for D_o) and C_{p3} (core-to-SG) are measured by LCR meter at 100-kHz test frequency, and they are also listed in Table 6.3.

Figure 6.21 shows the transformer winding structure. RM 8 core is employed, and single strand of enameled wire is adopted for these windings. The winding dimension and measured parameters are listed in Table 6.4, where, the primary winding turns number $N_P = 54$, and the secondary winding turns number $N_S = 10$.

The measured parameters of the transformer are listed in Table 6.5, where, L_{mp} is the magnetizing inductance and L_{lk_p} is the primary leakage inductance. The secondary leakage inductance is very small and can be neglected. Besides, C_{01} is the structural capacitance between W_{P1} and W_S, and C_{02} is the structural capacitance between W_{P2} and W_S, they are measured step-by-step during the manufacturing of the transformer. When W_{P1} and W_S winding layers have been wound, the measured structural capacitance of the transformer is C_{01}. After that, when W_{P2} has been wound, the measured structural capacitance is $C_{01} + C_{02}$, which is increased by C_{02}, so C_{02} can be calculated accordingly.

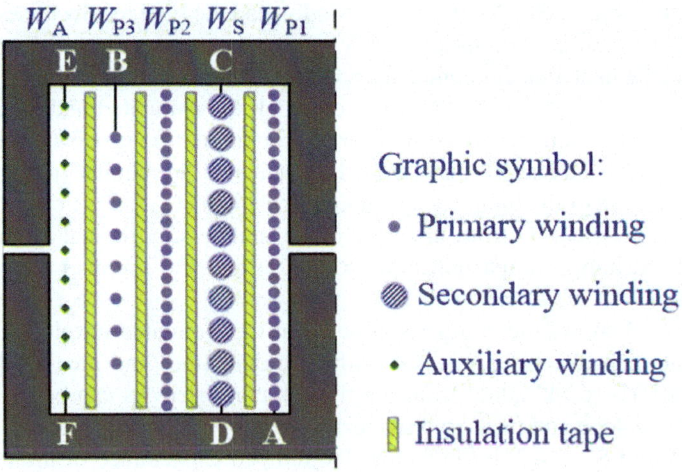

Fig. 6.21 The winding structure of the transformer

Table 6.4 The turns number and winding dimension of the transformer

Turns number		Winding dimension (mm)
Primary winding AB	$N_{P1} = 23\ (W_{P1})$	0.3
	$N_{P2} = 23\ (W_{P2})$	
	$N_{P3} = 8\ (W_{P3})$	
Secondary winding CD	$N_S = 10\ (W_S)$	0.5
Auxiliary winding EF	$N_A = 11\ (W_A)$	0.2

Table 6.5 The measured parameters of the transformer

Magnetizing inductance L_{mp}	715 μH	Structural capacitance C_{01}	25.50 pF
Leakage inductance L_{lk_p}	6.17 μH	Structural capacitance C_{02}	25.70 pF

Referring to (6.9) and (6.11), λ_P, λ_S and C_0 for the transformer in Fig. 6.21 are given by

$$\begin{cases} C_0 = C_{01} + C_{02} = 51.2\text{ pF} \\ \lambda_P = \frac{1}{C_0}\left(\frac{2N_P - N_{P1}}{2N_P} C_{01} + \frac{2N_P - 3N_{P1}}{2N_P} C_{02}\right) = 0.573 \\ \lambda_S = \frac{1}{C_0}\left(\frac{N_S + 0}{2N_S} C_{01} + \frac{0 + N_S}{2N_S} C_{02}\right) = 0.5 \end{cases} \tag{6.22}$$

2. The Experimental Results

The conducted EMI of the flyback converter is measured in the electromagnetic shielding chamber. In the test, the LISN employees R&S ENV216, the noise separator adopts Mini-Circuit ZSC-2-2 (Common Mode), and the EMI receiver uses Schaffner SCR3502. The test conditions are 220-V input rms voltage and full-load. Besides, the SG and the heat sink are connected to the PE in the measurement.

To verify the validity of the CM noise model, the simulated and measured CM noise spectra are compared. The main circuit for simulation is shown in Fig. 6.20b, and the leakage inductance and lumped capacitance are shown in Fig. 6.22. As seen, the solution of the lumped capacitance goes with (C_{AC}, C_{AD}, C_{BC}, C_{BD}) = (0, $\lambda_P C_0$, $\lambda_S C_0$, $(1 - \lambda_P - \lambda_S)C_0$). In addition, the snubber circuit, leakage inductance, the calculated lumped capacitance and C_{p1}, C_{p2} and C_{p3} are also included in the simulation.

Figure 6.23 shows the comparison between the simulated envelope (every two adjacent noise peaks are connected with a straight line) and measured spectrum of the CM EMI (peak value) from the flyback converter. As can be observed, the simulated envelope and measured spectrum are almost matched below 2 MHz.

Referring to Sect. 6.3, by including the lumped capacitance of the transformer and the parasitic capacitance C_{p1}, C_{p2} and C_{p3}, the ENS v_{ENS_FLY1} and v_{ENS_FLY2} of the flyback converter when the rectifier diode is at the high side and low side are respectively given by

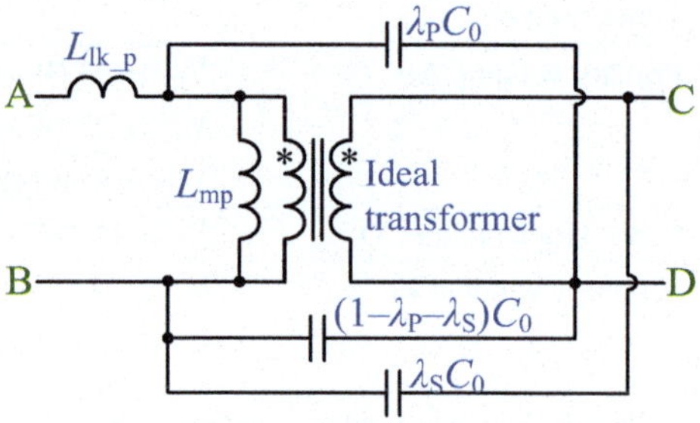

Fig. 6.22 The transformer parasitics in the simulation

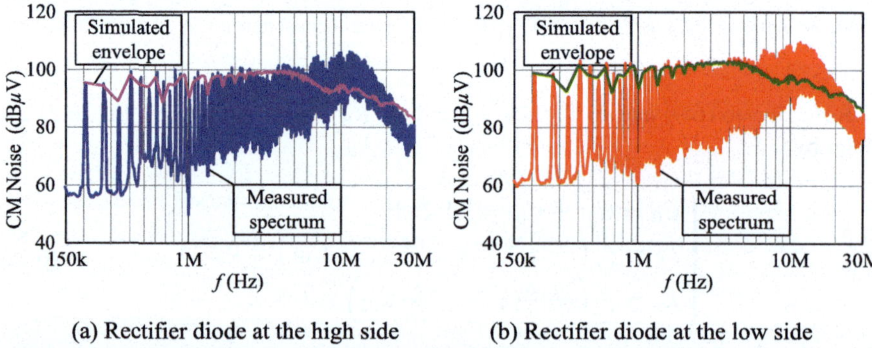

Fig. 6.23 The comparison between the measured spectrum and simulated envelope (peak)

$$v_{\text{ENS\_FLY1}} = \frac{(1 - \lambda_P - n\lambda_S)C_0 + C_{p1}}{C_0 + C_{p1} + C_{p2} + C_{p3}} v_Q \tag{6.23}$$

$$v_{\text{ENS\_FLY2}} = \frac{[1 - \lambda_P + n(1 - \lambda_S)]C_0 + C_{p1} + nC_{p2}}{C_0 + C_{p1} + C_{p2} + C_{p3}} v_Q \tag{6.24}$$

According to the CM noise model in Fig. 6.8b, the CM noise v_{CM} is related to the source impedance and the ENS of the converter. Under these two circuit configurations, the source impedance is the same, and the ENS is proportional to v_Q, so the CM noise spectrum is proportional to the coefficient before v_Q in the ENS. According to (6.23) and (6.24), the spectrum difference Δ_{FLY} is calculated, and it is expressed in decibles, i.e.,

$$\Delta_{FLY} = 20\lg\left|\frac{v_{\text{ENS\_FLY1}}}{v_{\text{ENS\_FLY2}}}\right| = 20\lg\left|\frac{(1 - \lambda_P - n\lambda_S)C_0 + C_{p1}}{[1 - \lambda_P + n(1 - \lambda_S)]C_0 + C_{p1} + nC_{p2}}\right| \text{(dB)} \tag{6.25}$$

By substituting λ_P, λ_S, C_0, C_{p1}, C_{p2}, C_{p3} and n into (6.25), Δ_{FLY} is equal to −2.75 dB. This means the CM noise spectrum of the flyback converter with rectifier diode at the high side is 2.75 dB lower than that with the rectifier diode at the low side.

The measured CM EMI spectra with different diode positions are provided in Fig. 6.24a. As can be seen, their CM noise spectra have similar pattern. Besides, the CM noise when the rectifier diode is at the high side is 2–3 dB lower than that when the rectifier diode is at the low side. As shown in Fig. 6.24b, by moving up the CM noise spectra when the rectifier diode is at the high side by 2.75 dB, the measured results are almost overlapped in the low frequency range from 150 kHz to 1 MHz. Therefore, the effectiveness of the ENS method in the low frequency range is verified.

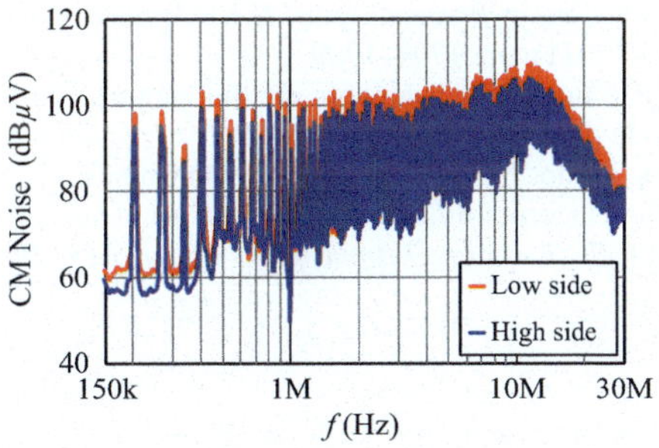

(a) Original case

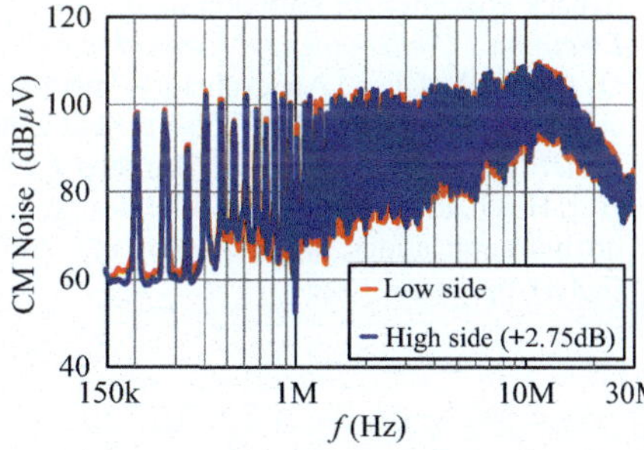

(b) Moving up the CM noise spectra (high side) by 2.75 dB

Fig. 6.24 The comparison of the measured spectrum with different position of D_o (peak)

6.6 Summary

In this chapter, a generalized lumped capacitance model for the transformer is proposed, which equivalently represents the inter-winding capacitance of the transformer with a general winding structure. With this model, the CM noise models of the basic isolated DC-DC converters, which consist of the source impedance and the equivalent noise source (ENS) voltage connected in series, are established for analyzing the CM noise. The source impedance equals to the transformer structural capacitance puls the CM parasitic capacitance between the high dv/dt nodes in the primary circuit and the PE. The ENS is related to the transformer winding structure and the circuit configuration. By analyzing the characteristics of the ENS, it is revealed that some topologies have the CMNC capability, including the flyback converter with output rectifier diode at the high side, the forward converter with rectifier diode and output filter inductor at the low side, as well as push–pull and half-bridge converters. For the full-bridge converter, it exhibits CMNC capability with PWM control, but lacks CMNC capability with phase-shift control. Finally, the lumped capacitance model for the transformer and the CM noise model of the isolated DC-DC converter are verified by the experimental results, respectively.

References

1. Y. Chu, S. Wang, A generalized common mode current cancellation approach for power converters. IEEE Trans. Industr. Electron. **62**(7), 4130–4140 (2015)
2. P. Kong, F.C. Lee, Transformer structure and its effects on common mode EMI noise in isolated power converters, in *Proceedings of IEEE Applied Power Electronics Conference and Exposition (APEC)* (2010), pp. 1424–1429
3. J. Collins, An accurate method for modeling transformer winding capacitance, in *Proceedings of Annual Conference of IEEE Industrial Electronics Society (IECON)* (1900), pp. 1094–1099
4. P. Meng, J. Zhang, H. Chen, Z. Qian, Y. Shen, Characterizing noise source and coupling path in flyback converter for common-mode noise prediction, in *Proceedings of IEEE Applied Power Electronics Conference and Exposition (APEC)* (2011), pp. 1704–1709
5. Q. Chen, W. Chen, An evaluation method of transformer behaviors on the suppression of common-mode conduction noise in switch mode power supply, in *Proceedings of the Chinese Society for Electrical Engineering (CSEE)*, vol. 32, no. 18 (2012), pp. 73–79 (in Chinese)
6. H.-I. Hsieh, D.-Y. Chen, EMI filter design method incorporating mix-mode conducted noise for off-line applications, in *Proceedings of IEEE Vehicle Power and Propulsion Conference (VPPC)* (2008), pp. 1617–1622

Chapter 7
Shielding-Based Common-Mode Noise Cancellation Techniques for Isolated DC-DC Converters

7.1 Shielding Techniques

In isolated DC-DC converters, the transformer inter-winding capacitance between the primary and secondary windings is a critical propagation path for common-mode (CM) noise. To block the displacement current flowing through this propagation path for reducing the CM noise, shielding layer is usually added between the primary and secondary windings. In this section, the performance of single- and double-shielding techniques will be reviewed.

7.1.1 Single-Shielding Technique

Figure 7.1a shows the single-shielding technique in wire-wound transformers, where, a shielding layer S_P is inserted between the primary-winding layer W_{P2} and the secondary-winding layer W_S, and it is connected to the primary ground (PG). With this shielding layer, the displacement current between W_{P2} and W_S is eliminated. However, two displacement currents are generated, which flow between S_P and W_{P2}, and between S_P and W_S, respectively. Figure 7.1b shows the propagation path for these two displacement currents in isolated DC-DC converters, where, the isolated DC-DC converters consist of the primary circuit, transformer, and the secondary rectifier and filter. The 25-Ω resistor is the equivalent impedance for the CM noise at the line impedance stabilization network (LISN) side.

According to Fig. 7.1b, the displacement current flowing between S_P and W_{P2} returns to the PG via the grounded wire of S_P and does not cause CM noise, while the current i_{dis} that flows between S_P and W_S returns to the PE and still causes the CM noise.

© The Author(s), under exclusive license to Springer Nature Singapore Pte Ltd. 2024

X. Ruan et al., *Conducted Electromagnetic Interference in Power Converters: Modeling, Prediction and Reduction*, CPSS Power Electronics Series,
https://doi.org/10.1007/978-981-99-9295-9_7

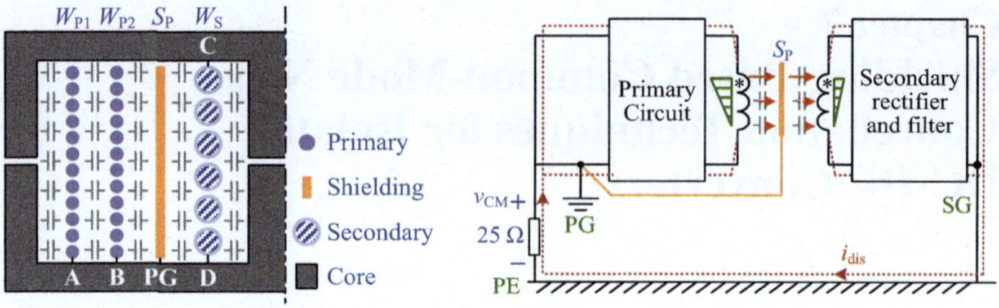

(a) The transformer with single shielding

(b) The CM noise propagation path

Fig. 7.1 The Transformer with single shielding and its CM noise propagation path

In Fig. 7.1, if S_P is connected to the secondary ground (SG), the displacement current flowing between S_P and W_S returns to the SG and does not cause CM noise, while the current that flows between S_P and W_{P2} returns to the PE and still causes the CM noise.

7.1.2 Double-Shielding Technique

In order to eliminate the displacement current flowing between S_P and W_S and enhance the CM noise reduction, another shielding layer S_S can be added between S_P and W_S, and it is connected to the SG, as shown in Fig. 7.2a. The technique with two shielding layers is referred to as the double-shielding technique [1].

Figure 7.2b shows the CM noise propagation path with the double-shielding technique. According to the cut set from the inter-winding capacitance between S_P and S_S to the PE, the displacement current flowing through S_P and S_S equals to i_{dis} that returns to the 25-Ω resistor. Since S_P and S_S have the same winding direction, the

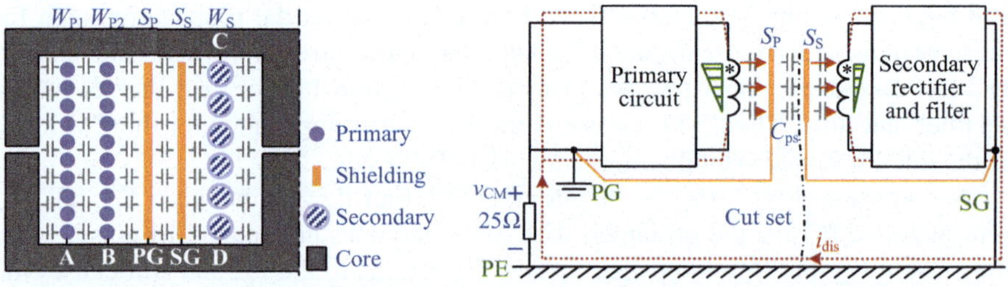

(a) The transformer with double shielding

(b) The CM noise propagation path

Fig. 7.2 The transformer with double-shielding and its CM noise propagation path voltage across the inter-winding capacitance C_{ps} between S_P and S_S equals to the voltage across the PG and SG. Thus, the S_P and S_S have the same electric potential distribution, and the CM noise is reduced. However, for the transformer with interleaved winding structure, two shielding layers should be added between every adjacent primary- and secondary-winding layers. As a result, these shielding layers will occupy large window and increase the complexity of manufacture.

7.2 Condition and Solutions to Achieve the CM Noise Cancellation

7.2.1 Condition to Achieve the CM Noise Cancellation

For the single-shielding technique, in order to achieve the common-mode noise cancellation (CMNC) capability, the displacement current flowing between the shielding layer and its adjacent secondary winding should be eliminated.

According to Sect. 6.2.2 in Chap. 6, the displacement current i_{dis} flowing between S_P and W_S in Fig. 7.1a is given by

$$i_{dis} = C_0 \frac{d}{dt}(\overline{v}_{SP} - \overline{v}_{WS}) \tag{7.1}$$

where, C_0 is the structural capacitance between S_P and W_S, and $\overline{v}_{SP}$ and $\overline{v}_{WS}$ are the averaged electric potential (AEP) of S_P and W_S with respect to the PE, respectively.

According to (7.1), the condition for $i_{dis} = 0$ is

$$\overline{v}_{SP} = \overline{v}_{WS} \tag{7.2}$$

Referring to Fig. 7.1b, $i_{dis} = 0$ indicates $v_{CM} = 0$, so the PG, PE and SG are equipotential in the conducted electromagnetic interference (EMI) frequency range. In this case, the AEP of S_P with respect to the PG is equal to that of W_S with respect to the SG. For the convenience of illustration, the reference ground for the shielding layer and secondary-winding layer is reselected as the PG and SG, respectively.

7.2.2 The AEP of Secondary-Winding Layers and Shielding Layers

1. A General Expression for the AEP of Secondary-Winding Layers

Figure 7.3a, b shows the typical structure of secondary-winding layer W_S with N_S turns in the wire-wound transformer and planar transformer, respectively. The electric

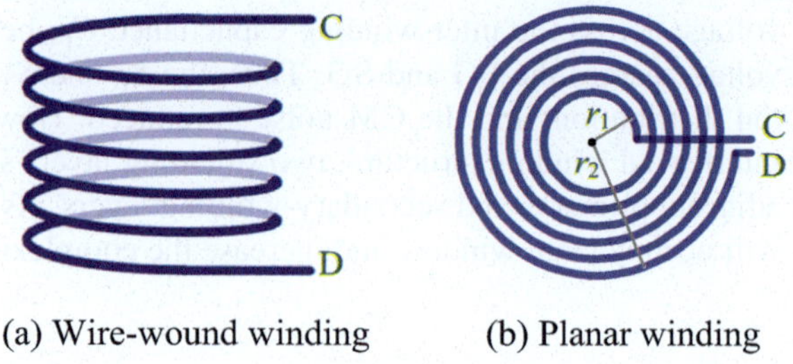

Fig. 7.3 The winding in wire-wound and planar transformer

potential of terminals C and D with respect to the SG are denoted by v_C and v_D, respectively.

According to Sect. 6.2.2 in Chap. 6, the AEP for a wire-wound winding is

$$\overline{v}_{WS} = 0.5(v_C + v_D) \quad \text{(wire - wound)} \tag{7.3}$$

The AEP of a planar winding is given by [2]

$$\overline{v}_{WS} = \lambda_1 v_C + (1 - \lambda_1) v_D \quad \text{(planar)} \tag{7.4}$$

where, λ_1 is expressed as

$$\lambda_1 = \frac{\frac{2}{3}r_1 + \frac{1}{3}r_2 + \frac{r_2 - r_1}{6N_S^2}}{r_1 + r_2} \tag{7.5}$$

where, r_1 is the inner radius, and r_2 is the outer radius.

Denoting the voltage across a single-turn winding as v_{st}, the electric potential of C is

$$v_C = N_S v_{st} + v_D \tag{7.6}$$

Substitution of (7.6) into (7.3) and (7.4), respectively, yields

$$\begin{cases} \overline{v}_{WS} = 0.5 N_S v_{st} + v_D & \text{(wire-wound)} \\ \overline{v}_{WS} = \lambda_1 N_S v_{st} + v_D & \text{(planar)} \end{cases} \tag{7.7}$$

2. General Expression for the AEP of Shielding Layers

In a wire-wound transformer, the shielding layer takes the cylindrical shape with one turn and covers the window, and its cross section is shown in Fig. 7.4a. In a planar transformer, the shielding layer is shown in Fig. 7.4b. Usually, the shielding layer is connected to the primary circuit via terminal G. The angle from terminal H to terminal G is denoted by θ_0 ($0 \le \theta_0 \le 2\pi$), and the electric potential of terminal G

Fig. 7.4 The top view of shielding layer in wire-wound and planar transformer

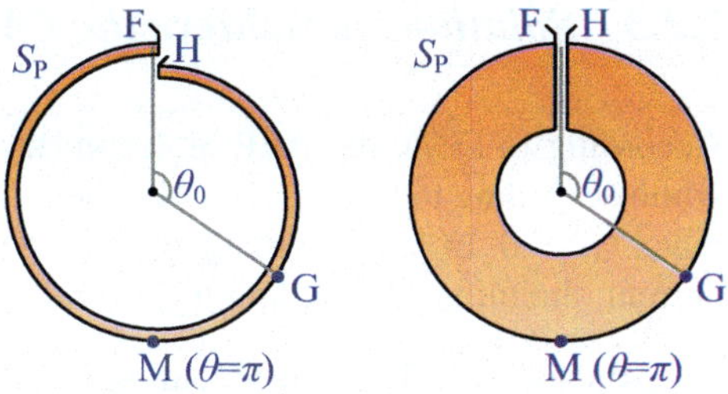

(a) Wire-wound transformer

(b) Planar transformer

with respect to the PG is denoted by v_G. Assuming that the electric potential of the shielding layer rises linearly from terminal H to terminal F, the electric potentials of terminals F and H are expressed as.

$$v_F = v_G + \frac{2\pi - \theta_0}{2\pi} v_{st} \tag{7.8}$$

$$v_H = v_G - \frac{\theta_0}{2\pi} v_{st} \tag{7.9}$$

According to (7.8) and (7.9), the AEP of shielding layers is expressed as

$$\overline{v}_{SP} = 0.5(v_F + v_H) = \frac{\pi - \theta_0}{2\pi} v_{st} + v_G \tag{7.10}$$

Since $\theta_0 \in [0, 2\pi]$, we have

$$\overline{v}_{SP} \in [-0.5v_{st} + v_G, 0.5v_{st} + v_G] \tag{7.11}$$

Normally, terminal G in the shielding layer is connected to the PG, so we have $v_G = 0$. Substituting $v_G = 0$ into (7.11), yields

$$\overline{v}_{SP} \in [-0.5v_{st}, 0.5v_{st}] \tag{7.12}$$

As seen, the AEP of shielding layers is within [–0.5v_{st}, 0.5v_{st}]. According to (7.7), the AEP of secondary windings will exceed [–0.5v_{st}, 0.5v_{st}] if the secondary winding has multiple turns and $v_D = 0$. In this case, the AEP of secondary-winding layer and shielding layer is never equal, and the CMNC capability of shielding technique can not be achieved.

7.2.3 *Solutions to Achieve the CMNC Capability*

According to (7.7), the AEP of secondary-winding layers depends on N_S and v_D, while according to (7.8) and (7.11), the AEP of shielding layers is only related to θ_0 when $v_G = 0$. In order to make the two AEP equal for cancelling the displacement current, the turns of shielding layers and v_G can be adjusted as follows.

1. Adjusting the Turns of Shielding Layers

When the secondary-winding layer W_S has turns of N_S, the shielding layer can be adjusted into a shielding winding with turns of N_{SP}. Take the wire-wound transformer as an example, the adjustement of shielding layer into shielding winding is shown in the left side in Fig. 7.5, where, the terminals of shielding winding is F and H, terminal G is connected to the PG, and $v_G = 0$.

Suppose the electric potential of S_P increases linearly from v_H, for wire-wound transformer and planar transformer, the AEP of S_P is given by

$$\begin{cases} \overline{v}_{SP} = 0.5(v_F + v_H) = 0.5[(v_H + N_{SP}v_{st}) + v_H] = 0.5N_{SP}v_{st} + v_H \quad \text{(wire-wound)} \\ \overline{v}_{SP} = \lambda_1 v_F + (1 - \lambda_1)v_H = \lambda_1(N_{SP}v_{st} + v_H) + (1 - \lambda_1)v_H = \lambda_1 N_{SP}v_{st} + v_H \quad \text{(planar)} \end{cases} \tag{7.13}$$

Substituting (7.7) and (7.13) into (7.1), i_{dis} for the wire-wound and planar transformers are

$$\begin{cases} i_{dis} = C_0\frac{d}{dt}[0.5v_{st}(N_{SP} - N_S) + (v_H - v_D)] \quad \text{(wire-wound)} \\ i_{dis} = C_0\frac{d}{dt}[\lambda_1 v_{st}(N_{SP} - N_S) + (v_H - v_D)] \quad \text{(planar)} \end{cases} \tag{7.14}$$

According to (7.14), in order to eliminate i_{dis}, $N_{SP} = N_S$ and $v_H = v_D$ should be satisfied. This means the S_P and W_S should have the same winding direction and turns, and terminals H and D at the same side should be equi-potential. As a result, the S_P and W_S have the same electric potential distribution. This method for cancelling the CM noise is called the shielding winding method. Since v_G in S_P is zero, there

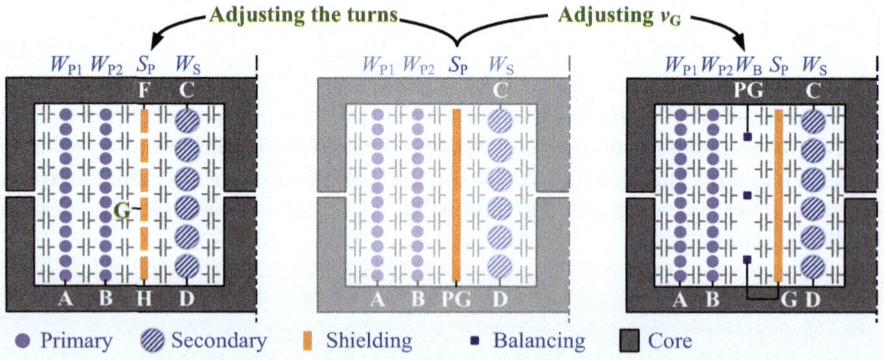

Fig. 7.5 Solutions to equalize the AEP of secondary-winding layer and shielding layer

exists a point in W_S whose electric potential is also zero. Based on this condition, the application of the shielding winding method in isolated DC-DC converters will be discussed in Sect. 7.3.

2. Adjusting v_G of Shielding Layers

v_G can be adjusted by connecting terminal G to an additional balancing winding W_B, as shown in the right side of Fig. 7.5, and the other terminal of W_B is connected to the PG. Note that W_B is placed between W_{P2} and S_P for avoiding the displacement current flowing through W_B and W_S. With W_B, v_G is changed as

$$v_G = \pm N_b v_{st} \tag{7.15}$$

where, N_b is the turns number of the balancing winding, and the sign ahead is determined by the relative direction of the balancing winding.

By substituting (7.15) into (7.10), the AEP of shielding layer with W_B is

$$\overline{v}_{SP} = \left(\frac{\pi - \theta_0}{2\pi} \pm N_b\right) v_{st} \tag{7.16}$$

For a wire-wound transformer, in order to make the AEP of S_P and W_S equal, the condition is derived by combining (7.2), (7.7) and (7.16), i.e.,

$$0.5 N_S v_{st} + v_D = \left(\frac{\pi - \theta_0}{2\pi} \pm N_b\right) v_{st} \quad \text{(wire-wound)} \tag{7.17}$$

According to (7.17), v_D is derived as

$$v_D = \left(\frac{\pi - \theta_0}{2\pi} \pm N_b - 0.5 N_S\right) v_{st} \quad \text{(wire-wound)} \tag{7.18}$$

Similarly, for a planar transformer, by combining (7.2), (7.7) and (7.16), we have

$$\lambda_1 N_S v_{st} + v_D = \left(\frac{\pi - \theta_0}{2\pi} \pm N_b\right) v_{st} \quad \text{(planar)} \tag{7.19}$$

According to (7.19), v_D is derived as

$$v_D = \left(\frac{\pi - \theta_0}{2\pi} \pm N_b - \lambda_1 N_S\right) v_{st} \quad \text{(planar)} \tag{7.20}$$

As seen from (7.18) and (7.20), if v_D is proportional to v_{st}, the balancing winding can be used to equalize the AEP of S_P and W_S. This method is called the combined shielding and balancing-winding method, and its applications in isolated DC-DC converters will be presented in Sect. 7.4.

7.3 Applications of the Shielding Winding Method

7.3.1 Combination of the Shielding Winding and Secondary Winding

The transformer usually adopts interleaved winding structure, so as to reduce the leakage inductance and AC resistance of the windings. In this case, the shielding winding should be inserted between every adjacent primary- and secondary-winding layers, as shown in Fig. 7.6a. Apparently, this will lead to complexity in manufacture. Noticed that S_{P1}, W_S and S_{P2} have the same turns and winding direction, these layers can be properly transformed and combined, as shown in Fig. 7.6b. The combined wire is defined as the shielding-secondary winding (SSW), as shown in Fig. 7.6c. The corresponding transformer structure implemented with the SSW is illustrated in Fig. 7.6d.

Based on the geometric structure of the SSW, the shielding layer can block the electric coupling between the center wire (secondary winding) and the other windings in the transformer. Since the shielding layer and the center wire have the same electric potential distribution, the displacement current between them is naturally eliminated regardless of the capacitance. Therefore, the SSW is quite easy and convenient. In practical applications, there are still some drawbacks and challenges when using

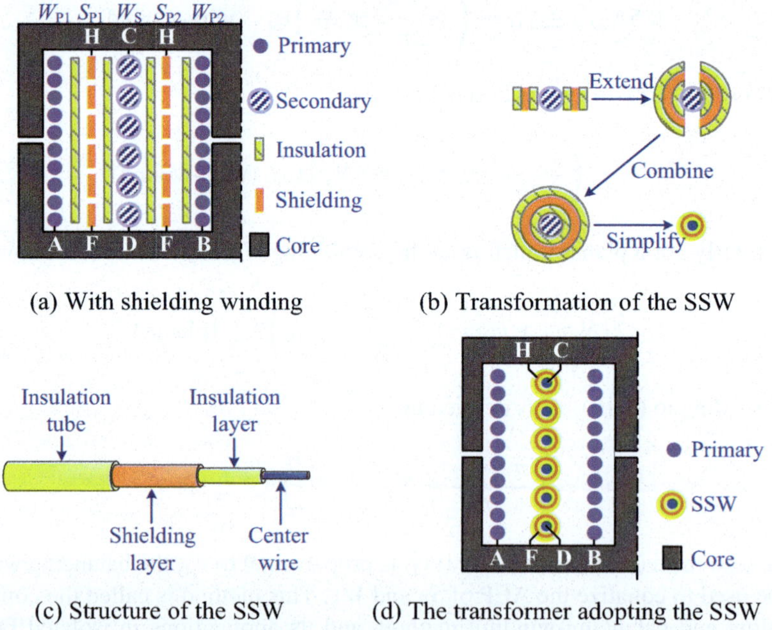

Fig. 7.6 The transformation and derivation of the SSW

the SSW: (1) The SSW occupies more window height than the enameled wire with the same diameter of the center wire; (2) The SSW will introduce larger leakage inductance of the transformer windings than the transformer without shielding layers; and (3) Since the center wire conducts relatively large current, the generated heat should be carefully handled.

7.3.2 Applications of the Shielding Winding Method

As mentioned in Sect. 7.2.3, when applying the shielding winding method, there should exist a point in W_S whose electric potential is zero. Such point is defined as the static-electric-potential (SEP) point, since the electric potential of these points with respect to the SG is not influenced by the switching action, such as the positive and negative terminals of the output voltage. In this Section, to investigate the application of the shielding winding method in basic isolated DC-DC converters, the existence of SEP point in secondary windings will be examined.

1. Condition for the Existence of SEP Point in Secondry Windings

There are two cases for the seconday winding with the SEP point. One is that the winding terminal is connected to the SEP point such as the positive or negative terminal of the output voltage, and the other is that the SEP point is inside the windings. In order to check the existence of SEP point inside the windings, the electric potential distribution along the secondary winding should be calculated first.

For a specific isolated converter and the corresponding operating mode, the electric potential of the secondary-winding terminals can be obtained. By neglecting the leakage magnetic flux and AC resistance caused by the skin effect and proximity effect, the electric potential varies linearly along the windings [2]. Figure 7.7a shows the transformer winding structure, where C and D are the origin and end terminals, respectively, and E is a point in the winding. The length from E to D along the winding is denoted by l_{ED}, and the whole winding length from C to D is denoted by l_{CD}. Based on the linear variation of the electric potential along the windings, the electric potential of E is given by.

$$v_E = v_D + \frac{v_C - v_D}{l_{CD}} l_{ED} = \frac{l_{ED}}{l_{CD}} v_C + \left(1 - \frac{l_{ED}}{l_{CD}}\right) v_D \tag{7.21}$$

According to (7.21), Fig. 7.7b depicts the curve of v_E.

Taking the derivative of (7.21) with respect to t, yields

$$\frac{\partial v_E}{\partial t} = \frac{l_{ED}}{l_{CD}} \frac{\partial v_C}{\partial t} + \left(1 - \frac{l_{ED}}{l_{CD}}\right) \frac{\partial v_D}{\partial t} \tag{7.22}$$

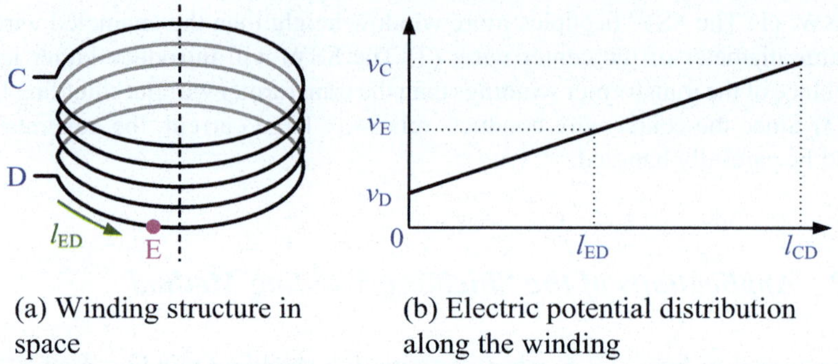

Fig. 7.7 The structure and electric potential distribution of the secondary winding

If point E is an SEP point, we have $\partial v_E/\partial t = 0$. According to (7.22), the condition for the existence of SEP point can be derived as

$$\frac{\partial v_C}{\partial t} = \frac{l_{ED} - l_{CD}}{l_{ED}} \frac{\partial v_D}{\partial t} \tag{7.23}$$

According to (7.23), l_{ED} can be solved to locate the position of point E.

$$l_{ED} = \frac{\frac{\partial v_D}{\partial t}}{\frac{\partial v_D}{\partial t} - \frac{\partial v_C}{\partial t}} l_{CD} \tag{7.24}$$

2. Applications of the Shielding Winding Method in Basic Isolated DC-DC Converters

Figure 7.8a shows the flyback converter. As seen, terminal D is connected to the SG, and it is the SEP point in the secondary winding. Therefore, the shielding winding can be used, where, the center wire is connected to terminals C and D, and the dotted terminal in the shielding winding is connected to the PG. Similar situations can be found in forward, push–pull and half-bridge converters, as shown in Figs. 7.8b–d.

Figure 7.9a shows the buck-type full-bridge converter with a full-wave rectifier. The secondary rectifier can also be the full-bridge and current-doubler rectifiers, as shown in Fig. 7.9b, c, respectively. For the full-wave rectifier shown in Fig. 7.9a, the center tap of the secondary windings is connected to the SG, and it is an SEP point, so the shielding winding method can be adopted.

Figure 7.10a shows the electric potential of terminals C and D of the secondary winding in the full-bridge rectifier. According to the waveforms of v_C and v_D, no SEP point exists in the secondary windings. Figure 7.10b shows the electric potential and its variation of the terminals in current-doubler rectifier. By comparing the electric potential variation of v_C and v_D with respect to t, there is no SEP point in the secondary winding. So, the shielding winding method can not be applied in the full-bridge rectifier and current-doubler rectifier.

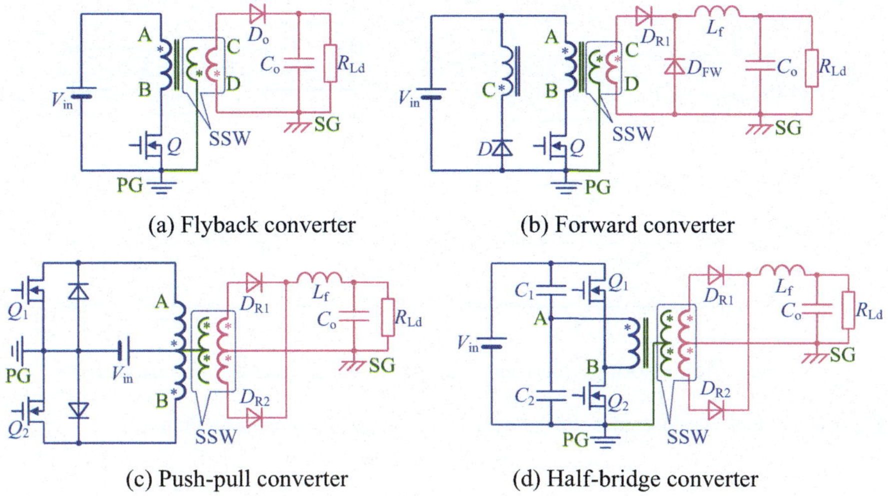

(a) Flyback converter
(b) Forward converter
(c) Push-pull converter
(d) Half-bridge converter

Fig. 7.8 The isolated DC-DC converter adopting the SSW method

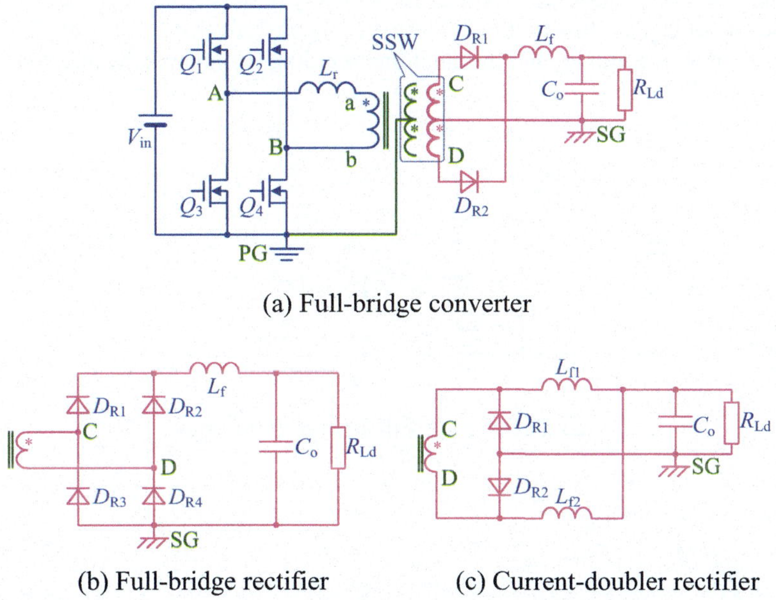

(a) Full-bridge converter
(b) Full-bridge rectifier
(c) Current-doubler rectifier

Fig. 7.9 The buck-type full-bridge converter and its secondary rectifier and filter

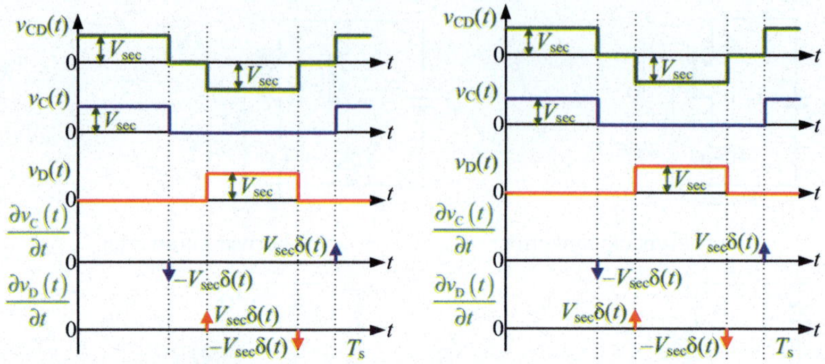

(a) The full-bridge rectifier (b) The current-doubler rectifier

Fig. 7.10 The electric potential and its variation of the terminals in the rectifiers

Figure 7.11 shows the boost-type full-bridge converter with a full-wave rectifier. Likewise, the secondary rectifier can also be the full-bridge and voltage-doubler rectifiers, as shown in Fig. 7.11b, c, respectively. For the full-wave and voltage-doubler rectifiers, as shown in Fig. 7.11a, b, it is obvious that there is an SEP point in the secondary winding, and the shielding winding method can be adopted.

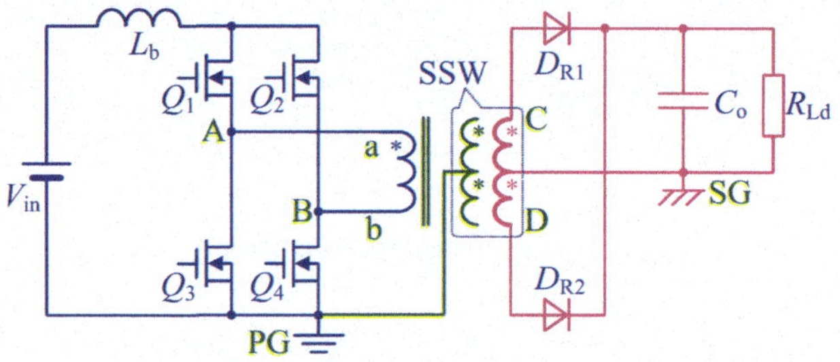

(a) Boost-type full-bridge converter

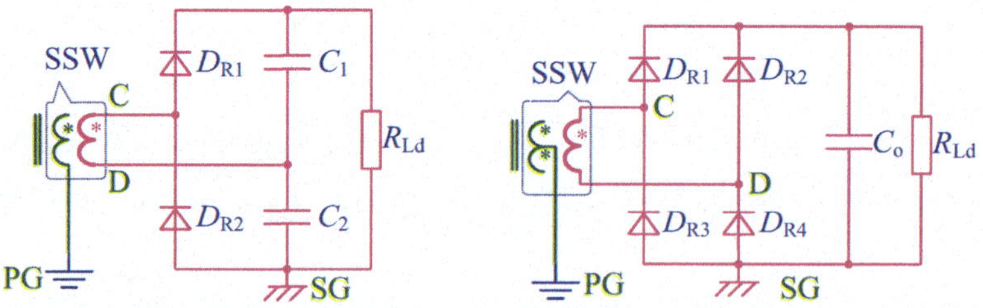

(b) Voltage-doubler rectifier (c) Full-bridge rectifier

Fig. 7.11 The boost-type full-bridge converter and its secondary rectifier and filters

For the full-bridge rectifier shown in Fig. 7.11c, the electric potential and its variation are given in Fig. 7.12. As seen, $\partial v_C / \partial t = -\partial v_D / \partial t$ is satisfied. According to (7.23), it can be obtained that there is an SEP point in the secondary winding. By substituting $\partial v_C/\partial t = -\partial v_D/\partial t$ into (7.24), we have $l_{ED} = l_{CD}/2$, which means the middle of the secondary winding is the SEP point. As shown in Fig. 7.11c, the center wire of the SSW is connected to the rectifier, and the middle of the shielding layer is connected to the PG.

Table 7.1 summarizes the existence of the SEP point in the transformer secondary windings in the basic isolated DC-DC converters. If there is an SEP point in the secondary winding, the shielding winding method can be applied to suppress the CM noise.

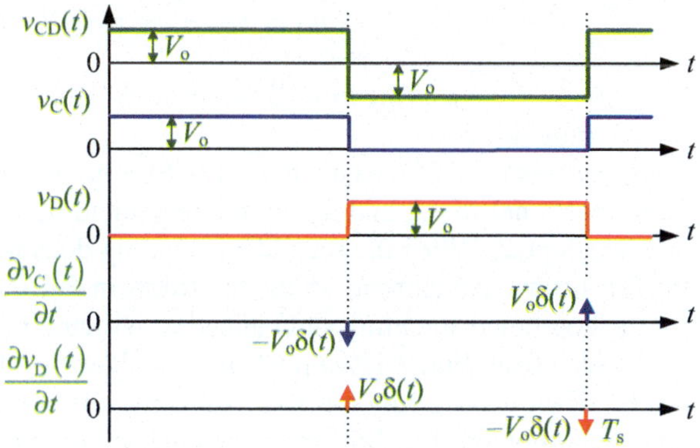

Fig. 7.12 The electric potential and its variation of the terminals in full-bridge rectifier

Table 7.1 Existence of SEP Point along the windings in basic isolated DC-DC converters

Topology		Existence of the SEP Point in the secondary winding and its connection
Flyback		Yes, the SEP point is connected to the SG
Forward		Yes, the SEP point is connected to the SG
Push–pull		Yes, the SEP point is connected to the center-tap
Half-bridge		Yes, the SEP point is connected to the center-tap
Buck-type full-bridge	Full-wave	Yes, the SEP point is connected to the center-tap of the secondary winding
	Full-bridge	No
	Current-doubler	
Boost-type full-bridge	Full-wave	Yes, the SEP point is connected to the SG
	Full-bridge	Yes, the SEP point is connected to the mid-point of the secondary winding
	Voltage-doubler	Yes, the SEP point is connected to the SG

7.4 Application of the Combined Shielding and Balancing-Winding Method

7.4.1 The Secondary Rectifier and Filters for Applying the Combined Shielding and Balancing-Winding Method

In Sect. 7.2.3, the condition for applying the combined shielding and balancing-winding method is that v_D is proportional to v_{st}. It is worth noting that, if there is an SEP point E in the secondary winding, we have

$$v_D = v_{DE} + v_E = \pm N_{DE} v_{st} \tag{7.25}$$

where, $v_E = 0$, N_{DE} is the turns from terminal D to point E, and the sign is related to the winding direction.

According to (7.25), when there is an SEP point in the secondary winding, v_D will be proportional to v_{st}, and the combined shielding and balancing-winding method can be adopted. When the secondary winding does not have the SEP point, such as the full-bridge and current-doubler rectifiers in Fig. 7.9b, c, the applicable condition for the combined shielding and balancing-winding method will be exaimed.

As seen from Fig. 7.10, v_{CD} has three voltage levels, so v_{st} also has three voltage levels. However, v_D only has two voltage levels, so v_D is never proportional to v_{st}, and the combined shielding and balancing-winding method can not applied.

According to the discussions above, the converters suitable for applying the shielding winding and the combined shielding and balancing-winding methods to reduce the CM noise are identical. Therefore, Table 7.1 can also be adopted to look up the applications of the combined shielding and balancing-winding method in basic isolated DC-DC converters.

7.4.2 Calculating the Turns of Balancing Winding and the Angle θ_0

By substituting (7.25) into (7.18) and (7.20), respectively, the requirement for eqalizing the AEP of the secondary winding and shielding layer is

$$\frac{\pi - \theta_0}{2\pi} \pm N_b = 0.5 N_S \pm N_{DE} \quad \text{(wire-wound)} \tag{7.26}$$

$$\frac{\pi - \theta_0}{2\pi} \pm N_b = \lambda_1 N_S \pm N_{DE} \quad \text{(planar)} \tag{7.27}$$

Noting that $(\pi-\theta_0)/(2\pi) \in [-0.5, 0.5]$ and N_b is an integer, therefore N_b and $(\pi-\theta_0)/2\pi$ are supposed to be equal to the integral and decimal part, respectively.

$$\begin{cases} N_b = \text{floor}(|0.5N_S \pm N_{DE}|) \\ \theta_0 = \pi - 2\pi(0.5N_S \pm N_{DE} \pm N_b) \end{cases} \quad \text{(wire-wound)} \tag{7.28}$$

$$\begin{cases} N_b = \text{floor}(|\lambda_1 N_S \pm N_{DE}|) \\ \theta_0 = \pi - 2\pi(\lambda_1 N_S \pm N_{DE} \pm N_b) \end{cases} \quad \text{(planar)} \tag{7.29}$$

where, floor(x) is the function which outputs the greatest integer less than or equal to x.

According to (7.28), for a wire-wound transformer, if the turns number N_S of a secondary winding is an even number, then $N_b = |0.5N_S \pm N_{DE}|$, and $\theta_0 = \pi$; Otherwise, $N_b = |0.5(N_S-1) \pm N_{DE}|$, and $\theta_0 = 0$. For a planar transformer, when $N_S = 1$, then $\lambda_1 = 0.5$ according to (7.5). Based on (7.29), $N_b = N_{DE}$, and $\theta_0 = 0$. If terminal D is an SEP point, then $N_b = N_{DE} = 0$, which is the shielding technique proposed in [3].

7.5 Hybrid Passive Cancellation Method

For isolated dc-dc converters with a symmetrical configuration, such as two-switch flyback, two-switch forward, push–pull, full-bridge and full-bridge *LLC* resonant converters, when a proper control scheme is adopted, complementary dv/dt nodes appear in their primary circuits, and the displacement current flowing through the related parasitic capacitances can be cancelled by each other. In this case, the inter-winding capacitance is the dominant propagation path for the CM noise, and both the shielding winding method and the combined shielding and balancing-winding method can reduce the CM noise effectively.

However, for asymmetrical isolated DC-DC converters, such as flyback, forward and half-bridge converters, the parasitic capacitance between the high dv/dt node in the primary circuit and the PE is another propagation path for the CM noise. In this section, the combined shielding and balancing-winding method is further combined with the passive cancellation method [4], which is called hybrid passive cancellation (HPC) method, will be investigated for cancelling the displacement current flowing through two propagation paths for the CM noise simultaneously.

7.5.1 *Principle of the HPC Method*

To cancel the displacement current flowing through the parasitic capacitance between the high dv/dt node in the primary circuit and the PE, the passive cancellation method was proposed by adding a compensation voltage source and a compensation capacitor

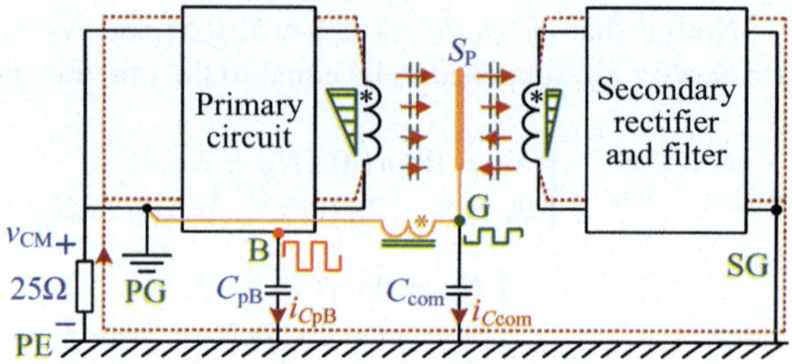

Fig. 7.13 The principle of the HPC method

[4], where, the compensation voltage source is realized by a compensation winding. For the combined shielding and balancing-winding method, node G in the shielding layer is a high dv/dt node. If the dv/dt of node G and the node in the primary circuit are out-of-phase, the balancing winding can be reused as the compensation winding, and the HPC method is proposed, as shown in Fig. 7.13, where, C_{com} is the compensation capacitor, node B is a high dv/dt node in the primary circuit, and C_{pB} is the parasitic capacitance between node B and the PE.

In Fig. 7.13, for canceling the displacement current i_{CpB} flowing through C_{pB}, the compensation current i_{Ccom} should be equal to $-\ i_{CpB}$, i.e.,

$$C_{pB}\frac{dv_B}{dt} + C_{com}\frac{dv_G}{dt} = 0 \tag{7.30}$$

As seen, the applicable condition for the HPC method can be summarized as follows:

(1) The electric potential of node D in the secondary winding is proportional to v_{st}, so that the combined shielding and balancing-winding method can be adopted to cancel the displacement current flowing between the shielding layer and the secondary-winding layer.
(2) The dv/dt of the primary node and node G of the balancing winding are out-of-phase, so that the balancing winding can be reused as the compensation winding.

In practical applications, a transformer may have several shielding layers. To equalize the AEP of shielding layers and their adjacent secondary windings, several balancing windings should be added, and thus several nodes between the shielding layers and balancing windings exist. The compensation capacitor can be connected to the node whose dv/dt is complementary with the node in the primary circuit. In addition, the combination of the shielding winding method and the passive cancellation method is also possible for reducing the CM noise, which reuses the shielding winding as the compensation winding, and it is not discussed here.

7.5.2 Applications of the HPC Method

1. Flyback Converter

Figure 7.14 shows the applications of the HPC method in flyback, forward, half-bridge and half-bridge *LLC* resonant converters. For the flyback converter in Fig. 7.14a, the AEP of secondary winding is linearly correlated with the electric potential of node C since $v_D = 0$. To eqaulize the AEP of the secondary winding and shielding layer, the dotted terminal of the balancing winding is connected to the PG. As a result, the electric potential of nodes G and C are in-phase. Since the electric potential of the drain of switch Q and node G in the shielding layer are in-phase, the balancing winding can not be reused as the compensation winding, and the HPC method is not applicable here.

For the flyback converter in Fig. 7.14b, the AEP of the secondary winding is linearly correlated with the electric potential of node D since $v_C = 0$. To equalize the AEP of the secondary winding and shielding layer, the dotted terminal of the balancing winding is connected to the shielding layer. Noting that the electric potential of the drain of switch Q and node G in the shielding layer are out-of-phase, the balancing winding can be reused as the compensation winding, and the HPC method can be applied. By neglecting the leakage inductance of the transformer, the electric potential of v_B and v_G satisfies

$$\frac{\mathrm{d}v_B}{\mathrm{d}t} = -\frac{N_P}{N_b} \cdot \frac{\mathrm{d}v_G}{\mathrm{d}t} \tag{7.31}$$

where, N_P is the turns of the primary winding.

Substitution of (7.31) into (7.30), leads to

$$C_{com} = \frac{N_P}{N_b} \cdot C_{pB} \tag{7.32}$$

2. Forward Converter

Figure 7.14c, d show the forward (FWD) converter, and the output rectifier diode D_{R1} and filter inductor L_f can be located at the high side or the low side. For the circuit shown in Fig. 7.14a, node G in the balancing winding and node C in the secondary winding are dotted terminals, so that the AEP of shielding layer and secondary winding can be equal. In addition, the electric potential of node G and the drain of switch Q are out-of-phase, and the HPC method is applicable. The corresponding C_{com} is the same as (7.32).

In Fig. 7.14b, to equalize the AEP of the shielding layer and secondary winding, node G in the balancing winding and node D in the secondary winding are dotted terminals. However, since the electric potential of the drain of switch Q and node G are in-phase, the balancing winding can not be reused as compensation winding, and the HPC method is not applicable.

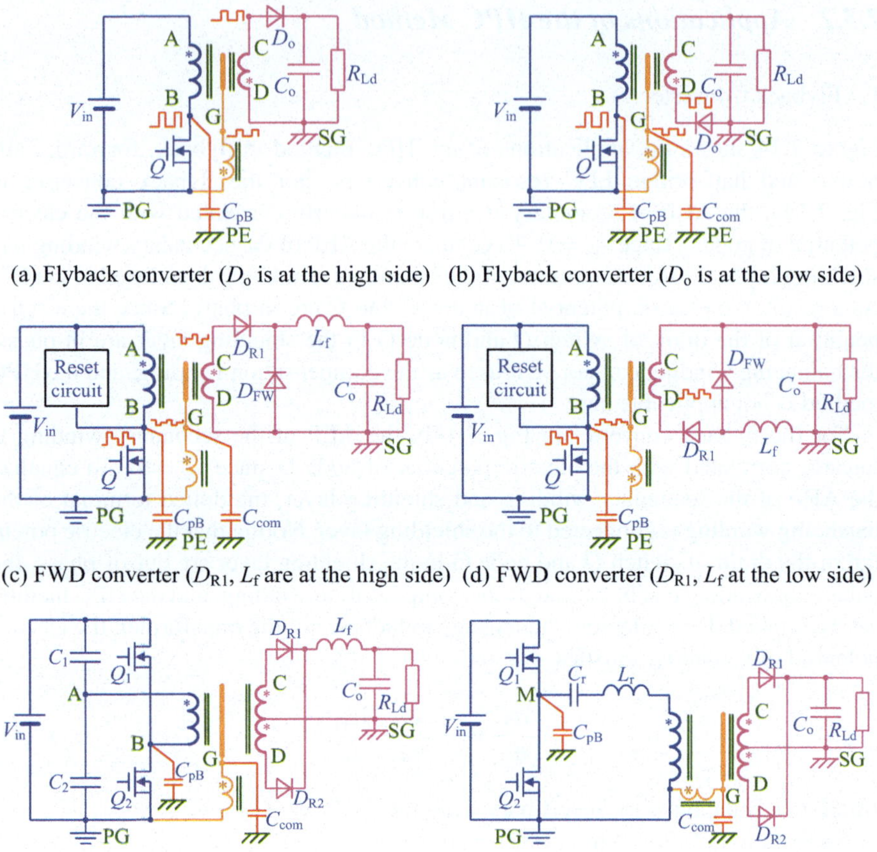

(a) Flyback converter (D_o is at the high side) (b) Flyback converter (D_o is at the low side)

(c) FWD converter (D_{R1}, L_f are at the high side) (d) FWD converter (D_{R1}, L_f at the low side)

(e) Half-bridge converter (f) Half-bridge *LLC* resonant converter

Fig. 7.14 The applications of the HPC method in flyback, forward, half-bridge and half-bridge *LLC* resonant converters

3. Half-Bridge Converter

Figure 7.14e shows the half-bridge (HB) converter using the HPC method. Node G in the balancing winding and node C in the secondary winding are dotted terminals, so that the AEP of the shielding layer and the secondary winding between node C and SG are equal. As a result, the electric potential of node G and the drain of switch Q_2 are out-of-phase, and the HPC method can be applied. The required C_{com} is the same as (7.32).

4. Half-Bridge *LLC* Resonant Converter

Figure 7.14f shows the HB *LLC* resonant converter adopting the HPC method. When the switching frequency and resonant frequency are close, the impedance of L_r and C_r in series is quite lower than that of the magnetizing inductance of the transformer.

Therefore, the voltage across the primary winding is approximately equal to the voltage between the phase-leg mid-point M and the PG. Therefore, the balancing winding can be reused as the compensation winding just like the HB converter, and C_{com} is the same as (7.32).

7.6 Experimental Verification and Discussion

In this section, the shielding winding method, the combined shielding and balancing-winding method, and the HPC method are verified by the experimental results, respectively.

7.6.1 Verification of the Shielding Winding Method

In order to verify the effectiveness of the shielding winding method, a 60-W prototype of the flyback converter has been built in the lab. Figure 7.15 shows the topology of the flyback converter. The input voltage v_{ac} is 90~264 V, and the output voltage V_o is 19.5 V. The devices used in the prototype are: main switch Q adopts AOTF10N65 (10A/650V), output rectifier diode D_o adopts V30120C (30 A/120 V), input filter capacitor employs 100-μF electrolytic capacitor, and the output filter capacitance is 1000 μF. A 0.47-μF X cap is inserted in front of the rectifier bridge, which filters the high frequency pulsating input current and eliminates the mixed-mode noise [5]. The converter is controlled by FAN6754MRMY with jitter frequency function to reduce the spectra of the noise source.

As the heatsink accommodating switch Q is connected to the PG, the dominant CM noise propagation path is the transformer inter-winding capacitance. To verify the effectiveness of the shielding winding method, three transformers are designed: $T_{1\text{-}1}$ without shielding layers, $T_{1\text{-}2}$ with single-shielding layers, and $T_{1\text{-}3}$ implemented with the SSW wires, as shown in Fig. 7.16.

Table 7.2 lists the winding dimension of these transformers. For the secondary winding in $T_{1\text{-}3}$, the diameter of the coaxial cable is 2.4 mm; the external and inner diameter of the shielding layer is 2.00 mm and 1.4 mm, respectively, and the diameter

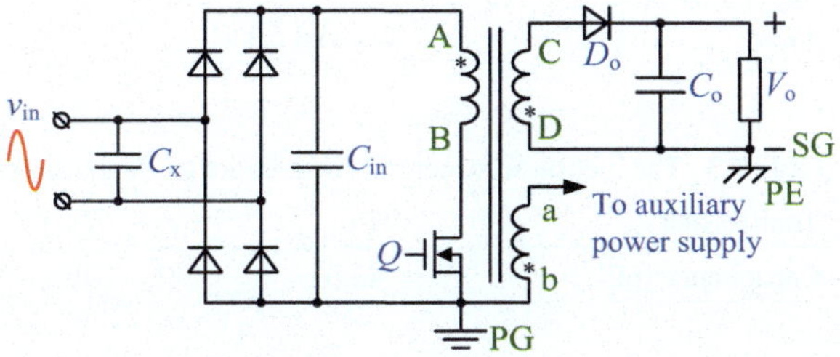

Fig. 7.15 The flyback converter and the connection of transformer

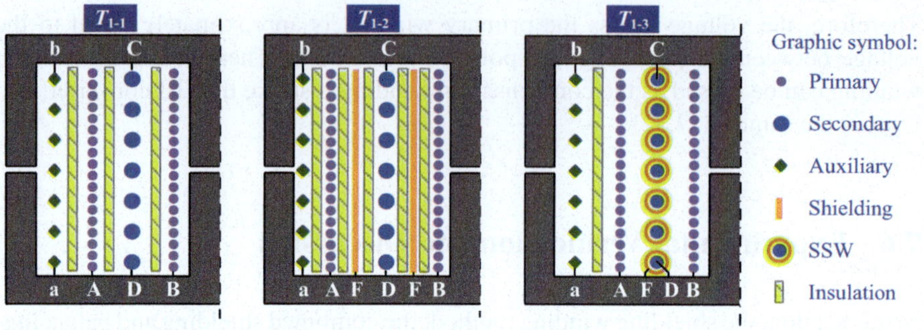

Fig. 7.16 The winding structure of three transformers

of the center wire is 0.45 mm. The transformers adopt PQ 32/30 ferrite core, and the magnetizing inductance of the primary winding is 460 μH.

Since the center wire of the coaxial cable is very thin, the test condition is 220-V input voltage and 20-W load for the converter with these transformers. For a fair comparison, the inter-winding capacitances of these transformers are kept very close by adjusting the insulation layers of each transformer. Table 7.3 lists the measured structural capacitances.

The conducted EMI of the flyback converter is measured in the electromagnetic shielding chamber. The LISN employes R&S ENV216, the EMI receiver uses Schaffner SCR3502, and the noise separator adopts Mini-Circuit ZSC-2-2 (Common Mode). The SG of the converter is connected to the PE. Besides, each core of these transformers is wrapped by copper foil which is connected to the PG, for blocking the electric coupling between the primary winding and the PE via the core. The measured CM noise of the converter using three different transformers is illustrated in Fig. 7.17. As seen, the quasi-peak (QP) CM noise of the converter with $T_{1\text{-}3}$,

Table 7.2 The winding dimension of three transformers

<table>
<tr><th></th><th>Primary winding</th><th>Secondary winding</th><th>Auxiliary winding</th><th>Shielding layer</th></tr>
<tr><td>$T_{1\text{-}1}$</td><td rowspan="3">0.25 mm enameled wire with 2 strands, and each layer has 18 turns</td><td rowspan="2">0.45 mm enameled wire with 7 strands, and has 7 turns</td><td rowspan="3">0.3 mm enameled wire with 1 strand, and has 7 turns</td><td></td></tr>
<tr><td>$T_{1\text{-}2}$</td><td>Copper foil with a width of 15.5 mm and thickness of 0.1 mm</td></tr>
<tr><td>$T_{1\text{-}3}$</td><td>Coaxial cable</td><td></td></tr>
</table>

Table 7.3 The measured structural capacitance of three transformer

Transformer	$T_{1\text{-}1}$	$T_{1\text{-}2}$	$T_{1\text{-}3}$
Capacitance (pF)	43.03	49.71	48.83

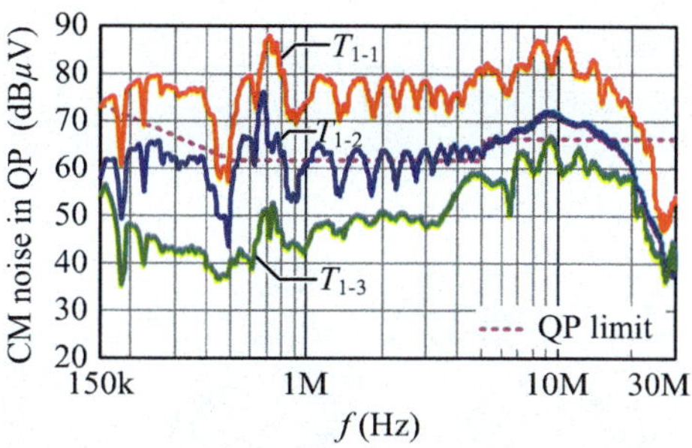

Fig. 7.17 The measured CM noise of the converter with three different transformers

which utilizes the SSW wire, is nearly 30 dB and 15 dB lower than $T_{1\text{-}1}$ than $T_{1\text{-}2}$, respectively. Therefore, the effectiveness of the shielding winding method is verified.

7.6.2 Verification of the Combined Shielding and Balancing-Winding Method

To verify the effectiveness of the combined shielding and balancing-winding method, the transformer without shielding layer, the transformer with shielding layer and the transformer with shielding layer and balancing winding are fabricated, and the displacement current flowing through the inter-winding capacitance is measured and compared [6].

Figure 7.18a shows the connection of the transformer under test. In order to establish the electric potential distribution of the windings, the primary winding is connected to the excitation port, which is an internal frequency-sweeping source in the network analyzer. A 10-kΩ resistor R is connected between terminal C and PG [6]. The voltage v_{AB} across the primary winding and v_{R} across R are sensed by the probes that connect the sensing ports of the network analyzer, and the transfer function $G_{\text{TR}}(s)$ from v_{AB} to v_{R} is measured as

$$G_{\text{TR}}(s) = \frac{V_{\text{R}}(s)}{V_{\text{AB}}(s)} \tag{7.33}$$

Figure 7.18b shows the equivalent circuit of the transformer under test. As seen, the current i_{shs} flowing from the shielding layer to the secondary winding is equal to the current i returning to the PG. Since v_{R} is proportional to the amplitude of i, the magnitude of $G_{\text{TR}}(s)$ can be used to evaluate the displacement current flowing from the shielding layer to the adjacent secondary winding.

Figure 7.19 shows the winding structure of three planar transformers, including $T_{2\text{-}1}$ without shielding layer, $T_{2\text{-}2}$ with shielding layer, and $T_{2\text{-}3}$ with shielding layer and balancing winding. For the combined shielding and balancing-winding method in $T_{2\text{-}3}$, S_{P1} and S_{P2} are specially given at the bottom of Fig. 7.19. Table 7.4 lists

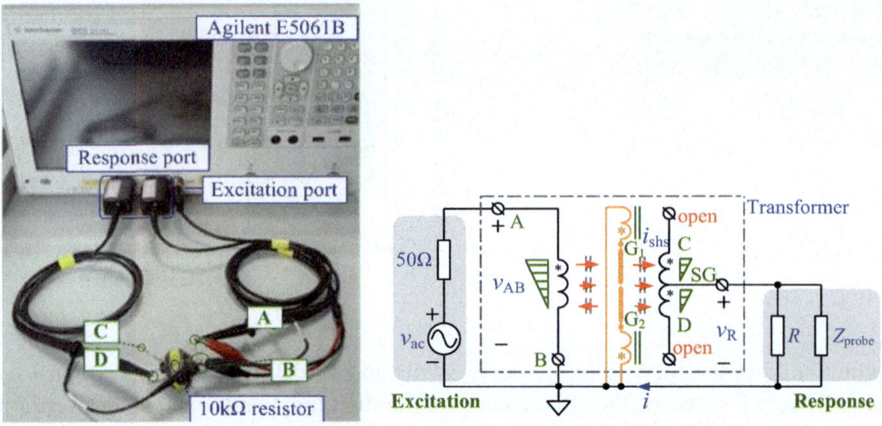

(a) The connection of the transformer under test

(b) The equivalent circuit

Fig. 7.18 The test of the transformer and the corresponding equivalent circuit

the winding dimension of these transformers. PQI 50/33 ferrite core is used for these planar transformers, and the magnetizing inductance is 98 μH for the primary winding. The measured structural capacitances between Pri1 and Sec1, Pri2 and Sec 4 in $T_{2\text{-}1}$ are 19 pF and 23 pF, respectively, and the measured structural capacitance between S_{P1} and Sec1, S_{P2} and Sec 4 in $T_{2\text{-}2}$ and $T_{2\text{-}3}$ are 37 pF and 34 pF, respectively.

Figure 7.20 shows curves of the magnitude of $G_{TR}(s)$ for these transformers. As seen, 15-dB reduction is achieved after inserting the shielding layer, and 5-dB reduction is further achieved by adding the balancing winding. Therefore, the displacement current flowing through the shielding layer and secondary winding is reduced.

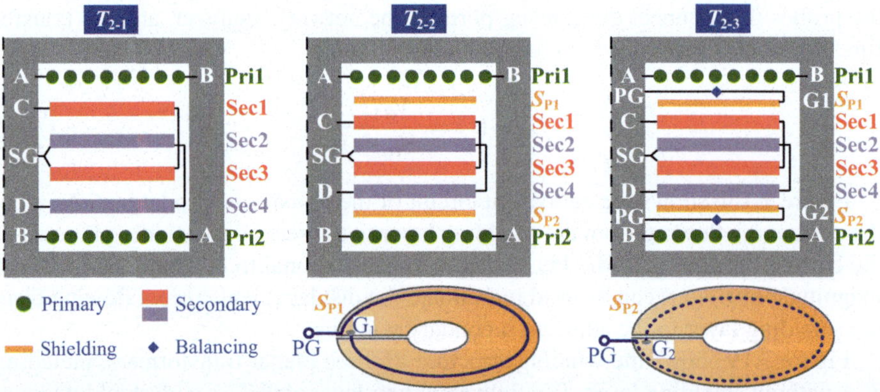

Fig. 7.19 The planar transformer under test and the corresponding winding structure

Table 7.4 The winding dimensions of planar transformer

	Wire type	Dimension	Turns
Primary	Litz wire	0.1 mm × 80	W_{P1}: 8
			W_{P2}: 8
Secondary	PCB winding	4 oz, 9.75 mm	4
Shielding	Copper foil	0.1 mm	
Balancing	Single strand enameled wire	0.25 mm	1

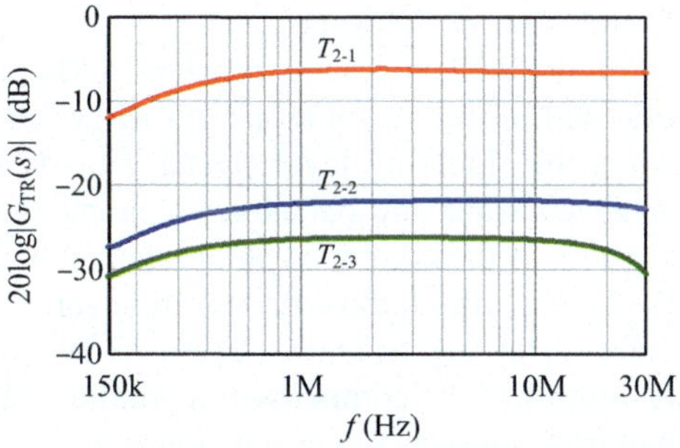

Fig. 7.20 The measured magnitude curves of three planar transformers

7.6.3 Verification of the HPC Method

To verify the effectiveness of the HPC method, a HB *LLC* resonant converter is built, as shown in Fig. 7.21. The related parameters are listed in Table 7.5, and the transformer uses the planar transformer shown in Fig. 7.19. In addition, the resonant inductor uses 150 × 0.1 mm litz wire with 10 turns in a PQ 26/20 core.

The measured parasitic capacitance C_{p1} and C_{p2} between the drain of Q_1, Q_2 and the heat sink is 17 pF. When measuring the CM EMI, the heat sink and SG is connected to the PE. The LISN uses 21115-50-TS-100-N from Solar Electronics,

Fig. 7.21 Photo of the prototype

Table 7.5 The specifications of the prototype

Input voltage V_{in}	385 V	Resonant inductance L_r	13.5 μH
Output voltage V_o	48 V	Resonant capacitance C_r	57 nF
Output power P_o	1 kW	Magnetizing inductance L_m	98 μH
Resonant frequency f_r	180 kHz	Input capacitance C_{in}	2.5 μF
Turns ratio n	1/4	Output capacitance C_o	600 μF

noise separator adopts ZSC-2-2 from Mini-Circuit and EMI receiver uses ESRP3 from Rohde & Schwarz.

Figure 7.22 shows the measured CM noise of the converter under full-load condition with different solutions. As shown in Fig. 7.22a, 10-dB reduction is achieved when the shielding layer (using $T_{2\text{-}2}$) is added. In Fig. 7.22b, 5-dB reduction is achieved when the balancing winding (using $T_{2\text{-}3}$) is added, verifying the effectiveness of the combined shielding and balancing-winding method. When applying the HPC method, the compensation capacitor is connected between the PE and the node G_2 in S_{P2}, as shown in Fig. 7.19. Because the turns of the primary winding is 4 times of the compensation winding (shielding layer + balancing winding), the required compensation capacitance is $4C_{p2}$, which is 68 pF. In Fig. 7.22c, by adding a 62-pF capacitor to achieve the best reduction, the CM noise is further reduced by 10 dB. Figure 7.22d compares the original CM noise and the CM noise with the HPC method, the total CM noise reduction is 25 dB, verifing the effectiveness of the HPC method.

7.7 Summary

In isolated DC-DC converters, shielding layer is usually added between the primary- and secondary-winding layers to block the displacement current and reduce the CM noise. In this chapter, it is pointed out that the displacement current flowing between the shielding layer and the adjacent secondary-winding layer still contributes to the CM noise. To eliminate this displacement current, it is necessary to equalize the AEP of shielding layer and the adjacent secondary-winding layer. To achieve equal AEP, two methods are proposed, including the shielding winding method and the combined shielding and balancing-winding method. The shielding winding method adjusts the turns of shielding layer, and the combined shielding and balancing-winding method adds a balancing winding to adjust v_G. Both of these methods can be applied in flyback, forward, push–pull, half-bridge, boost-type full-bridge converters and buck-type full-bridge converters with full-wave rectifiers. However, for the buck-type full-bridge converter with full-bridge rectifiers and current-doubler rectifiers, these methods can not be applied. To further eliminate the displacement current flowing

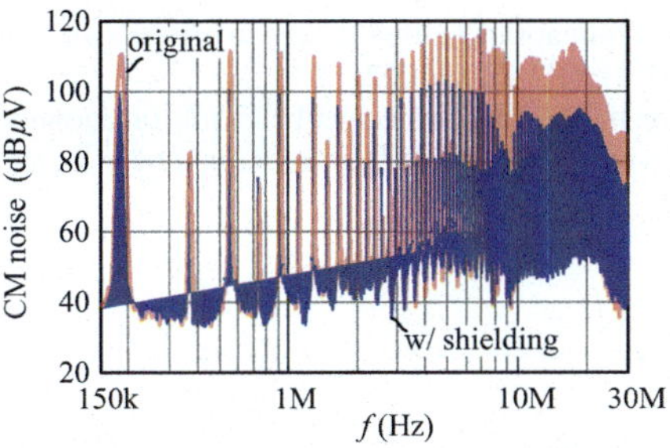

(a) Original case and with shielding

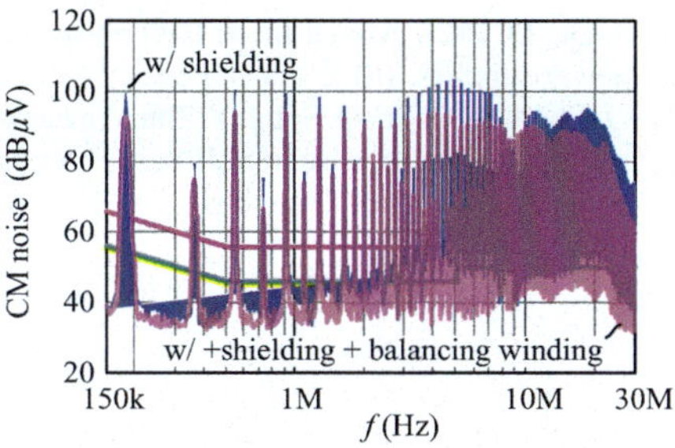

(b) With shielding and w/ shielding + balancing winding

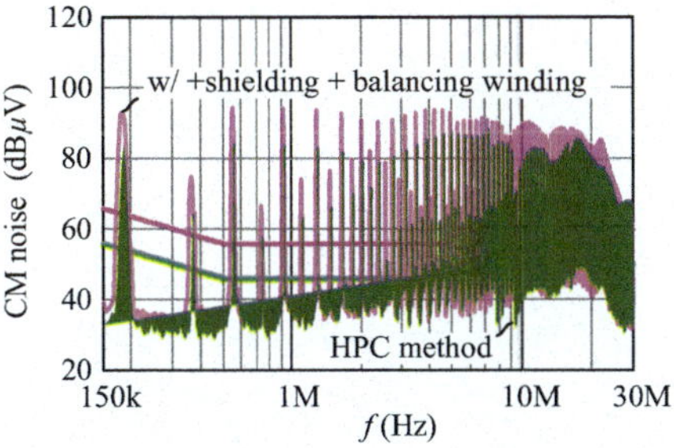

(c) With shielding + balancing winding and HPC method

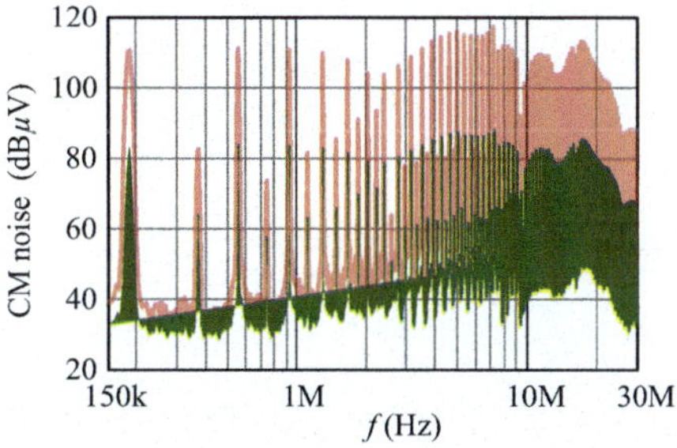

(d) Original case and HPC method

Fig. 7.22 Comparison of the CM noise in the half-bridge *LLC* resonant converter

through the high d*v*/d*t* node in the primary circuit and the PE, the HPC method is proposed. This method synthesizes the combined shielding and balancing-winding method and passive cancellation method, where the balancing winding is reused as a compensation winding. The shielding winding method, combined shielding and balancing-winding method and the HPC method are verified by the experimental results, respectively.

References

1. D.F. Knurek, Reducing EMI in switch mode power supplies, in *Proceedings of Telecommunication Energy Conference* (1988), pp. 411–420
2. P. Kong, S. Wang, F.C. Lee, Z. Wang, Reducing common-mode noise in two-switch forward converter. IEEE Trans. Power Electron. **26**(5), 1522–1533 (2011)
3. Y. Yang, D. Huang, F. C. Lee, Q. Li, Transformer shielding technique for common mode noise reduction in isolated converters, in *Proceedings of IEEE Energy Conversion Congress and Exposition (ECCE)* (2013), pp. 4149–4153
4. D. Cochrane, D. Chen, D. Boroyevic, Passive cancellation of common-mode noise in power electronic circuits. IEEE Trans. Power Electron. **18**(3), 756–763 (2003)

5. S. Qu, D. Chen, Mixed-mode EMI noise and its implications to filter design in offline switching power supplies. IEEE Trans. Power Electron. **17**(4), 502–507 (2002)
6. Q. Chen, W. Chen, Q. Song, Y. Zhu, An evaluation method of transformer behaviors on common-mode conduction noise in SMPS, in *Proceedings of IEEE Power Electronics and Drive Systems (PEDS)* (2011), pp. 782–786s

Chapter 8
Reducing the Common-Mode Noise in Phase-Shifted Full-Bridge Converter

8.1 The Common-Mode (CM) Noise Model of the Phase-Shifted Full-Bridge (PSFB) Converter

8.1.1 The PSFB Converter and Its CM Noise Model

As mentioned in Chaps. 6 and 7, the phase-shifted full-bridge (PSFB) converter does not have the common-mode (CM) noise cancellation capability, and the shielding-based methods can not reduce its CM noise. In this chapter, the CM noise reduction in the PSFB converter will be investigated.

Figure 8.1a shows the main propagation paths of the CM noise in the PSFB converter, where, $Q_1 \sim Q_4$ are the power switches, L_r is the resonant inductor, T_r is the isolation transformer with turns ratio of n, D_{R1} and D_{R2} are the output rectifier diodes, L_f is the output filter inductor, C_f is the output filter capacitor, and R_{Ld} is the load resistor. In addition, C_{ps} is the transformer inter-winding capacitance, C_{pA} and C_{pB} are the parasitic capacitance between the phase-leg midpoints A and B to the PE, respectively. When measuring the conducted EMI, the secondary ground (SG) is connected to the protective earth (PE), and the line impedance stabilization network (LISN) provides specified impedance in the conducted EMI frequency range from 150 kHz to 30 MHz.

Figure 8.1b illustrates the key waveforms of the PSFB converter, where, v_A and v_B are the voltages of the phase-leg midpoints with respect to the primary ground (PG), v_{Lr} is the voltage across the resonant inductor. During $[0, t_1]$ and $[t_3, t_4]$, D_{R1} and D_{R2} are both conducting, and the input voltage V_{in} is fully applied on L_r; during the other time intervals, the voltage applied on L_r is the difference between V_{in} and the secondary side voltage reflected to the primary, which is nearly zero.

© The Author(s), under exclusive license to Springer Nature Singapore Pte Ltd. 2024
X. Ruan et al., *Conducted Electromagnetic Interference in Power Converters: Modeling, Prediction and Reduction*, CPSS Power Electronics Series,
https://doi.org/10.1007/978-981-99-9295-9_8

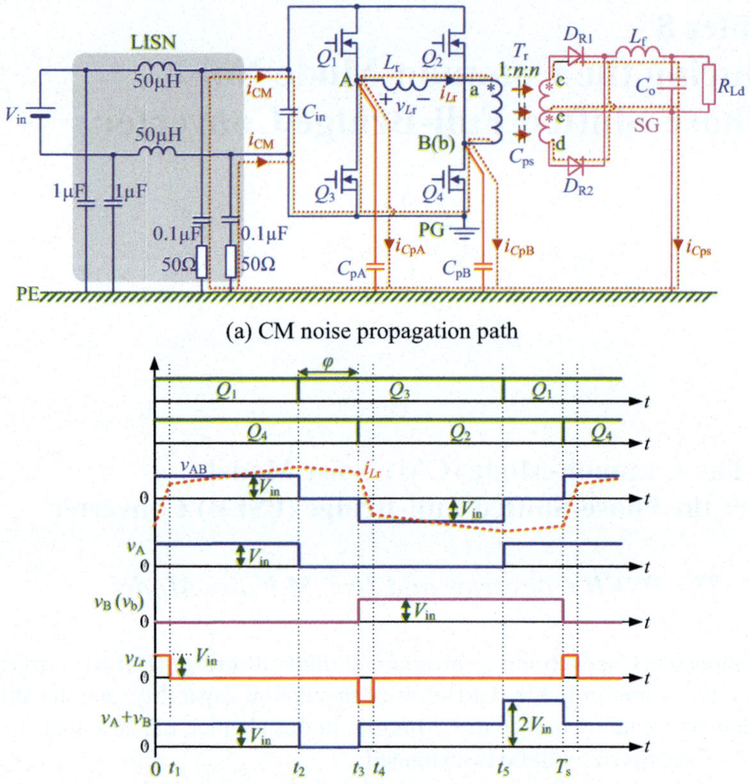

(a) CM noise propagation path

(b) Key waveforms with phase-shift control

Fig. 8.1 The CM noise propagation path and key waveforms of the PSFB converter

In Fig. 8.1a, the CM noise propagation paths are plotted with dashed lines. The displacement currents that flow through C_{pA} and C_{pB} are determined by the electric potentials of midpoints A and B, respectively; while the current flowing through C_{ps} is related to the electric potentials of the transformer winding terminals. As the resonant inductor voltage affects the electric potential of the transformer primary winding terminals, it also affects the CM noise. So, v_A, v_B and v_{Lr} are regarded as the voltage sources that generate the CM noise.

The substitution theorem is used for deriving the CM noise model of the full-bridge converter. The rule of substitution is to acquire a CM noise circuit that is easy to be analyzed, while avoiding voltage loops and current nodes [1]. Firstly, Q_3, Q_4 and L_r are replaced by the voltage sources with their own waveforms. In the frequency range from 150 kHz to 30 MHz, the impedance of the LISN for the CM noise is 25 Ω. Besides, the input capacitor C_{in} can be treated as short circuit owning to its relatively lower dv/dt compared with v_A, v_B or v_{Lr}. By neglecting the transformer leakage inductance, a controlled voltage/current source can be used to replace the transformer windings. To avoid voltage loops, Q_1, Q_2, D_{R1}, D_{R2} are replaced by

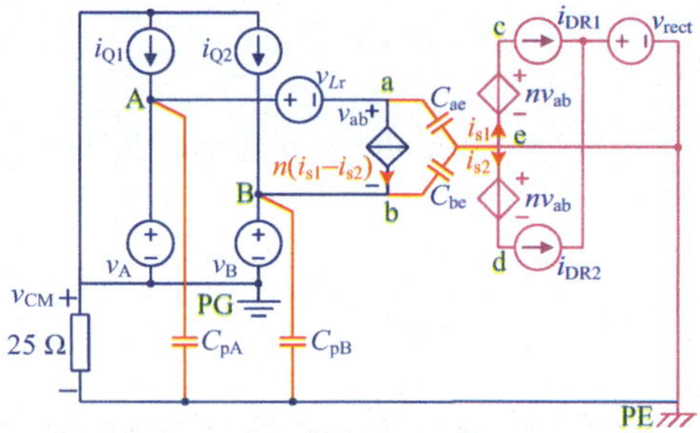

Fig. 8.2 The CM noise model of full-bridge converter with substituted sources

the current sources with their own current waveforms. Then, the branch consisted of the output filter (filter inductor and filter capacitor) and the load is replaced by a voltage source, in order to avoid current nodes. By applying the two-capacitor model proposed in [2], the transformer inter-winding capacitance is replaced by two capacitors C_{ae} and C_{be}, which will be derived in Sect. 8.1.2 by calculating the displacement. Finally, the CM noise model of the full-bridge converter is obtained, as shown in Fig. 8.2.

8.1.2 *The Derivation of the Two Capacitor C_{ae} and C_{be}*

Figure 8.3 shows the inter-winding capacitance model and the two-capacitor model of the transformer. As derived in Chap. 6, a general formula for the displacement current flowing through the inter-winding capacitance is given by

$$i_{dis} = C_0 \frac{d}{dt}\{\lambda_P v_a + (1-\lambda_P) v_b - [\lambda_S v_c + (1-\lambda_S) v_d]\} \tag{8.1}$$

where, C_0 is the transformer structural capacitance, λ_P and λ_S are transformer winding structural parameters.

Referring to Fig. 8.3b, the total displacement current that flows through C_{ae} and C_{be} is expressed as

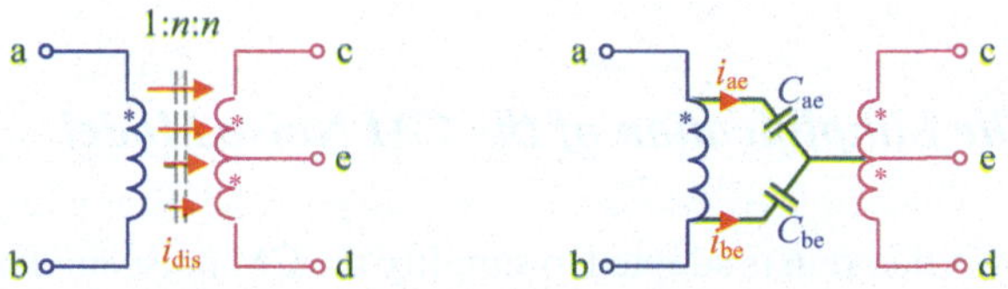

(a) Inter-winding capacitance model (b) Two-capacitor model

Fig. 8.3 The inter-winding capacitance and two-capacitor model of the transformer

$$i_{\mathrm{ae}} + i_{\mathrm{be}} = C_{\mathrm{ae}}\frac{\mathrm{d}}{\mathrm{d}t}(v_{\mathrm{a}} - v_{\mathrm{e}}) + C_{\mathrm{be}}\frac{\mathrm{d}}{\mathrm{d}t}(v_{\mathrm{b}} - v_{\mathrm{e}}) \tag{8.2}$$

By neglecting the transformer leakage inductance, the electric potential of the secondary winding terminals c and d are

$$\begin{cases} v_{\mathrm{c}} = n(v_{\mathrm{a}} - v_{\mathrm{b}}) + v_{\mathrm{e}} \\ v_{\mathrm{d}} = -n(v_{\mathrm{a}} - v_{\mathrm{b}}) + v_{\mathrm{e}} \end{cases} \tag{8.3}$$

Substituting (8.3) into (8.1), yields

$$i_{\mathrm{dis}} = C_{\mathrm{O}}\frac{\mathrm{d}}{\mathrm{d}t}[m v_{\mathrm{a}} + (1 - m)v_{\mathrm{b}} - v_{\mathrm{e}}] \tag{8.4}$$

where, the coefficient m is expressed as

$$m = \lambda_{\mathrm{P}} + n\ (1 - 2\lambda_{\mathrm{S}}) \tag{8.5}$$

Reorganizing (8.4) to approach the form of (8.1), yields

$$i_{\mathrm{dis}} = C_{\mathrm{O}}\frac{\mathrm{d}}{\mathrm{d}t}[m v_{\mathrm{a}} - m v_{\mathrm{e}} + (1 - m)v_{\mathrm{b}} - (1 - m)v_{\mathrm{e}}] = mC_{\mathrm{O}}\frac{\mathrm{d}}{\mathrm{d}t}v_{\mathrm{ae}} + (1 - m)C_{\mathrm{O}}\frac{\mathrm{d}}{\mathrm{d}t}v_{\mathrm{be}} \tag{8.6}$$

The condition for these two models to be equivalent is that the total displacement current flowing through the inter-winding capacitance is equal to that flowing through the two capacitors. By comparing the coefficient before v_{ae} and v_{be} in (8.6) and (8.2), we have

$$\begin{cases} C_{\mathrm{ae}} = mC_{\mathrm{O}} \\ C_{\mathrm{be}} = (1 - m)C_{\mathrm{O}} \end{cases} \tag{8.7}$$

Substituting (8.5) into (8.7), yields

$$\begin{cases} C_{\mathrm{ae}} = [\lambda_{\mathrm{P}} + n\ (1 - 2\lambda_{\mathrm{S}})]C_{\mathrm{O}} \\ C_{\mathrm{be}} = \{1 - [\lambda_{\mathrm{P}} + n\ (1 - 2\lambda_{\mathrm{S}})]\}C_{\mathrm{O}} \end{cases} \tag{8.8}$$

8.1.3 *The Simplification of the CM Noise Model*

Superposition theorem is adopted to simplify the CM noise model in Fig. 8.2. Firstly, only the excitation of current sources is considered, and all the voltage sources are shorted. The resultant sub-circuit is shown in Fig. 8.4a. As seen, all the current sources are essentially shorted, implying that current sources have no impact on the CM noise. Secondly, only the excitation of voltage sources is considered and all the

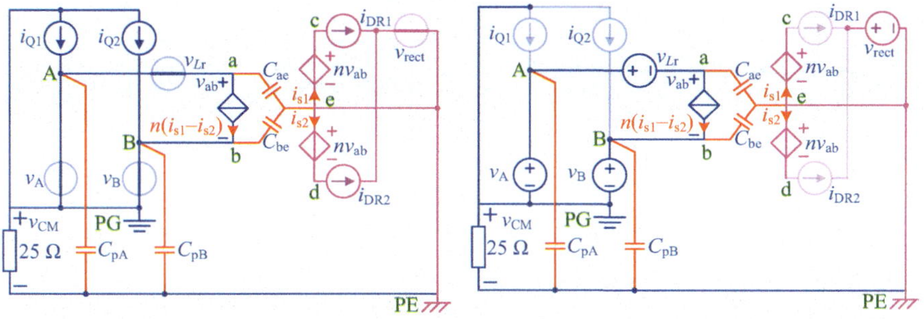

(a) Excited only by current sources (b) Excited only by voltage sources

Fig. 8.4 Simplification of CM noise model of full-bridge converter

Fig. 8.5 CM noise model of full-bridge converter

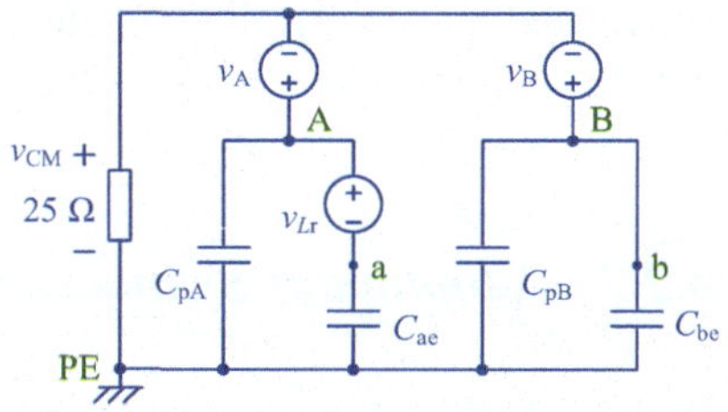

current sources are opened, and the resultant sub-circuit is shown in Fig. 8.4b. As seen, the transformer secondary winding currents, i_{s1} and i_{s2}, are both zero. Therefore, the controlled current source across the primary winding is essentially zero. Finally, the circuit in Fig. 8.4b is reorganized and simplified, and the CM noise model is finally obtained, as shown in Fig. 8.5. As seen, in order to reduce the CM noise in the PSFB converter, the displacement currents generated by v_{Lr}, v_A and v_B via the corresponding parasitic capacitance should be eliminated, which will be discussed in Sects. 8.2 and 8.3, respectively.

8.2 Applying Symmetry Circuit to Eliminate the Impact of Resonant Inductor Voltage on the CM Noise

8.2.1 Adoption of Symmetrical Resonant Inductor

As seen from Fig. 8.5a, C_{ae} is the only propagation path for v_{Lr}. In order to eliminate the effect of v_{Lr}, a compensation voltage source v'_{Lr} can be added and connected in series with C_{be}, as shown in Fig. 8.6a. Noted that v'_{Lr} should be complementary with v_{Lr}, and C_{be} should be equal to C_{ae}. By properly designing the transformer winding structure, the condition of $C_{be} = C_{ae}$ can be achieved. In addition, v'_{Lr} can

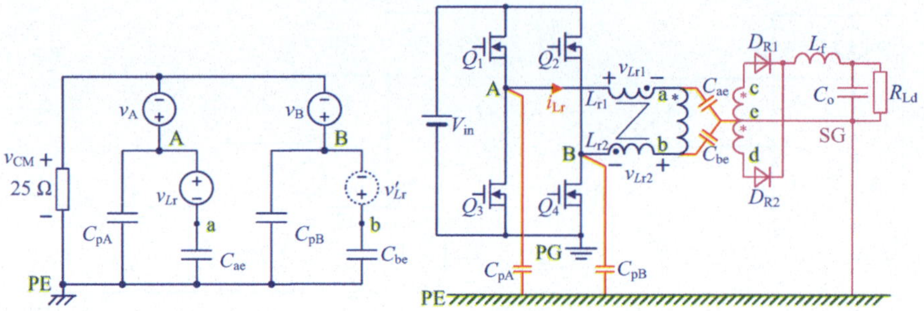

(a) The CM noise model with v'_{Lr} (b) The realization with symmetrical inductor

Fig. 8.6 CM noise model with compensation voltage source v'_{Lr}

be realized by equally dividing the resonant inductor winding into two ones, which are the coupled inductors L_{r1} and L_{r2}, as shown in Fig. 8.6b.

8.2.2 Adoption of Symmetrical Transformer

With $C_{\mathrm{ae}} = C_{\mathrm{be}}$, the propagation path of the transformer for the CM noise is symmetrical. Here, planar transformer is adopted to reduce the profile, as shown in Fig. 8.7. Spiral winding and PCB winding are used for the primary and secondary windings, respectively. The primary windings Pri1 and Pri2 are connected in series, and each winding has N_{P} turns. The secondary windings Sec1-1, Sec1-2, Sec2-1 and Sec2-2 are connected in series one by one, and each winding has one turn. Due to the non-evenly distributed inter-winding capacitance between the spiral windings, the terminal connection of the windings may influence the displacement current flowing through inter-winding capacitances [3]. Here, two kinds of terminal connection for the primary winding are investigated, as shown in Fig. 8.7.

Provided that Pri1 and Pri2 are tightly wound layers, there is displacement current flowing between Pri1 and Sec1-1, and between Pri2 and Sec2-2. The structural capacitance between Pri1 and Sec1-1 is denoted by C_{01}, and the structural capacitance between Pri2 and Sec2-2 is C_{02}. According to the displacement current formula given in [4], for terminal connection I, the displacement current i_{dis1} flowing from Pri1 to Sec1-1 and i_{dis2} flowing from Pri2 to Sec2-2 are expressed by

$$\begin{cases} i_{\mathrm{dis1}} = C_{01}\frac{\mathrm{d}}{\mathrm{d}t}\left[\lambda_1 v_{\mathrm{a}} + (1-\lambda_1)v_{\mathrm{M}} - \left(\frac{7}{8}v_{\mathrm{c}} + \frac{1}{8}v_{\mathrm{d}}\right)\right] \\ i_{\mathrm{dis2}} = C_{02}\frac{\mathrm{d}}{\mathrm{d}t}\left[\lambda_1 v_{\mathrm{M}} + (1-\lambda_1)v_{\mathrm{b}} - \left(\frac{1}{8}v_{\mathrm{c}} + \frac{7}{8}v_{\mathrm{d}}\right)\right] \end{cases} \tag{8.9}$$

where, v_{a}, v_{b}, v_{M}, v_{c} and v_{d} are the electric potentials of the corresponding transformer terminals, and λ_1 is a parameter depending on the winding dimension, which is given by

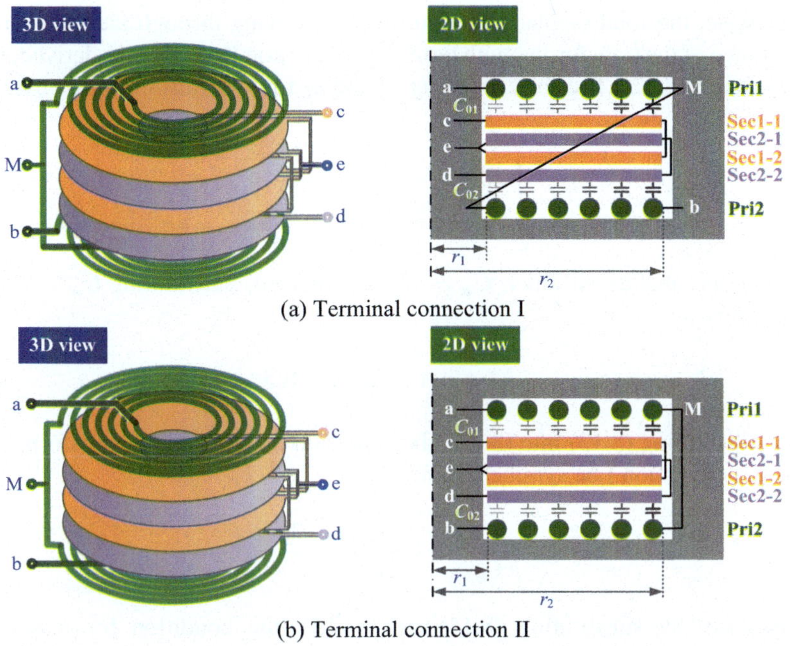

Fig. 8.7 Planaer transformer with different terminal connection of primary winding

$$\lambda_1 = \frac{\frac{2}{3}r_1 + \frac{1}{3}r_2 + \frac{r_2 - r_1}{6N_P^2}}{r_1 + r_2} \tag{8.10}$$

where, r_1 and r_2 denotes the inner and outer radius of windings.

Supposing that the electric potential varies linearly along windings, the electric potential of the midpoint M from the primary winding equals to

$$v_M = (v_a + v_b)/2 \tag{8.11}$$

By substituting (8.11) into (8.9), the total displacement current i_{dis} is derived as

$$\begin{aligned} i_{dis} = i_{dis1} + i_{dis2} &= (C_{01} + C_{02})\frac{d}{dt}[\lambda_P v_a + (1 - \lambda_P)v_b - \lambda_S v_c - (1 - \lambda_S)v_d] \\ &= C_0 \frac{d}{dt}[\lambda_P v_a + (1 - \lambda_P)v_b - \lambda_S v_c - (1 - \lambda_S)v_d] \end{aligned} \tag{8.12}$$

where, the transformer structural capacitance $C_0 = C_{01} + C_{02}$, and λ_P and λ_S are given by

$$\begin{cases} \lambda_P = \frac{(1+\lambda_1)C_{01} + \lambda_1 C_{02}}{2(C_{01} + C_{02})} \\ \lambda_S = \frac{7C_{01} + C_{02}}{8(C_{01} + C_{02})} \end{cases} \tag{8.13}$$

Likewise, the total displacement current i_{dis} flowing through the inter-winding capacitance of the transformer with terminal connection II can be also derived, which is the same as (8.12). The corresponding λ_P and λ_S for this transformer are expressed as

$$\begin{cases} \lambda_P = \frac{(1+\lambda_1)C_{01}+(1-\lambda_1)C_{02}}{2(C_{01}+C_{02})} \\ \lambda_S = \frac{7C_{01}+C_{02}}{8(C_{01}+C_{02})} \end{cases} \quad (8.14)$$

For a symmetrical transformer, $C_{ae} = C_{be}$ is expected. Substituting $C_{ae} = C_{be}$ into (8.8), yields

$$\lambda_P + n(1 - 2\lambda_S) = 0.5 \quad (8.15)$$

By substituting (8.13) into (8.15), the condition for making the transformer with terminal connection I be symmetrical is

$$\frac{C_{01}}{C_{02}} = \frac{2 - 2\lambda_1 - 3n}{2\lambda_1 - 3n} \quad (8.16)$$

Likewise, by substituting (8.14) into (8.15), the condition for making the transformer with terminal connection II be symmetrical is

$$2\lambda_1 - 3n = 0 \text{ or } C_{01} = C_{02} \quad (8.17)$$

To satisfy (8.16), C_{01} and C_{02} should be tuned, which is practically complex. While $C_{01} = C_{02}$ can be easily achieved to satisfy (8.17) by adopting the same type of insulation material with equal thickness. Therefore, terminal connection II is preferred.

From the above discussion, the symmetrical resonant inductor and symmetrical transformer configuration can be easily realized by adjusting the resonant inductor configuration and properly designing the transformer winding structure, which does not require additional cost.

8.3 Applying Passive Cancellation to Eliminate the Impact of Midpoint Voltages on the CM Noise

Usually, the same type of insulation material with the same thickness is inserted between $Q_1 \sim Q_4$ and the heat sink. So, C_{pA} and C_{pB} are nearly the same, i.e.,

$$C_{pA} = C_{pB} = C_p \quad (8.18)$$

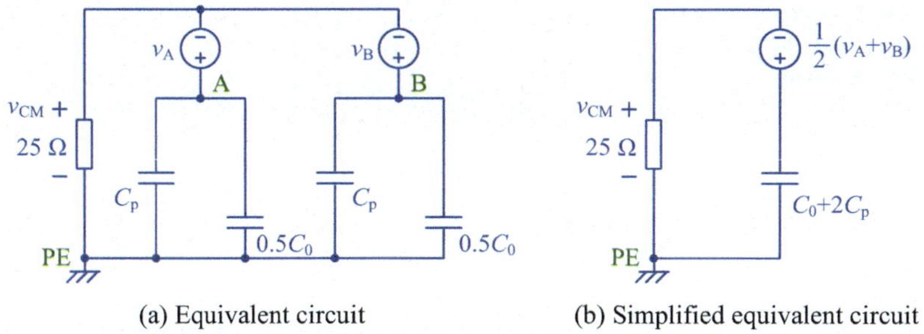

Fig. 8.8 CM noise model of full-bridge converter with symmetrical circuit

For a symmetrical transformer, the constraint of winding structural parameters is given in (8.15). By substituting (8.15) into (8.8), C_{ae} and C_{be} is derived as

$$C_{ae} = C_{be} = C_0/2 \tag{8.19}$$

With the symmetrical resonant inductor and symmetrical transformer, the effect of the resonant inductor voltage on the CM noise is nullified. Therefore, by shorting v_{Lr} and v'_{Lr} in Fig. 8.6a and considering the related capacitances given in (8.18) and (8.19), the CM noise model shown in Fig. 8.6a can be simplified, as shown in Fig. 8.8a.

Figure 8.8a can be further simplified as Fig. 8.8b with Thevenin's theorem. As seen, for cancelling the CM noise, a compensation branch consisting of a compensation voltage source v_{com} and a capacitor C_{com} can be added, as shown in Fig. 8.9. Moreover, v_{com} should be out-of-phase with ($v_A + v_B$), and the condition for cancelling the CM noise is

$$C_{com}\frac{d}{dt}v_{com} + (C_0 + 2C_p)\frac{d}{dt}\left(\frac{v_A + v_B}{2}\right) = 0 \tag{8.20}$$

According to (8.20), two implementations for realizing v_{com} will be presented as follows.

Fig. 8.9 CM noise model of full-bridge converter with cancellation branch

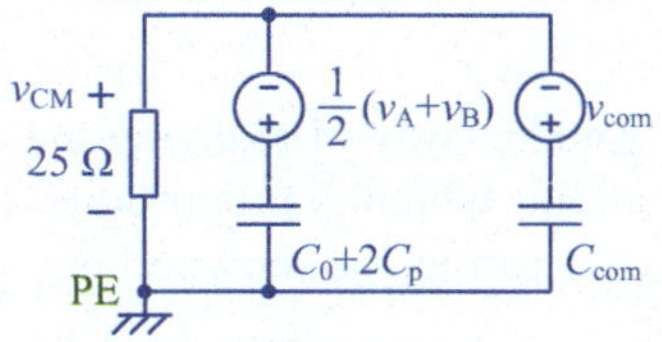

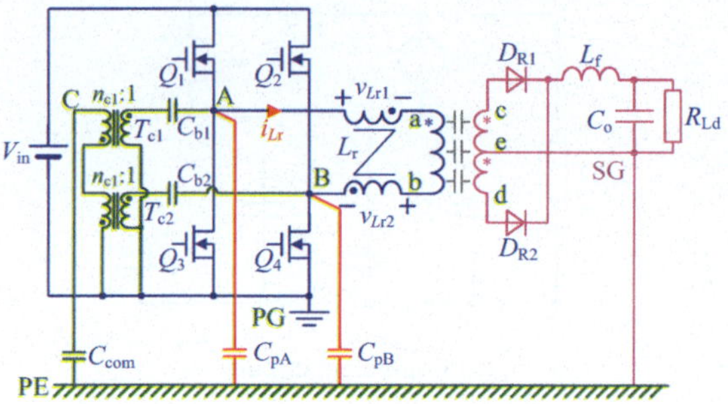

Fig. 8.10 Implementation I for constructing the cancellation circuit

8.3.1 Implementation I

To obtain the compensation voltage v_{com}, which is out-of-phase with ($v_A + v_B$), it is straightforward to sense v_A and v_B [5], and then invert them and add together, as shown in Fig. 8.10, where, C_{b1} and C_{b2} are the dc blocking capacitors for removing the DC component in v_A and v_B, which are nearly equal to $V_{in}/2$. The AC components of v_A and v_B are sensed by transformer T_{c1} and T_{c2}, respectively. The secondary sides of T_{c1} and T_{c2}, which output the corresponding inverted voltage, are connected in series for summing up the sensed voltages. The resultant compensation voltage v_{com} is

$$v_{com} = -n_{c1}(v_A + v_B - V_{in}) \tag{8.21}$$

where, n_{c1} is the turns ratio of T_{c1} and T_{c2}.

By substituting (8.21) into (8.20), the required compensation capacitance C_{com} is derived as

$$C_{com} = \frac{C_O + 2C_p}{2n_{c1}} \tag{8.22}$$

8.3.2 Implementation II

Another way of realizing the compensation voltage is to pick up the voltage in the circuit which is proportional to ($v_A + v_B$), and then invert the sensed voltage by a transformer, as shown in Fig. 8.11. With symmetrical resonant inductor, we have

$$\begin{cases} v_a = v_A - v_{Lr1} \\ v_b = v_B + v_{Lr2} \\ v_{Lr1} = v_{Lr2} \end{cases} \tag{8.23}$$

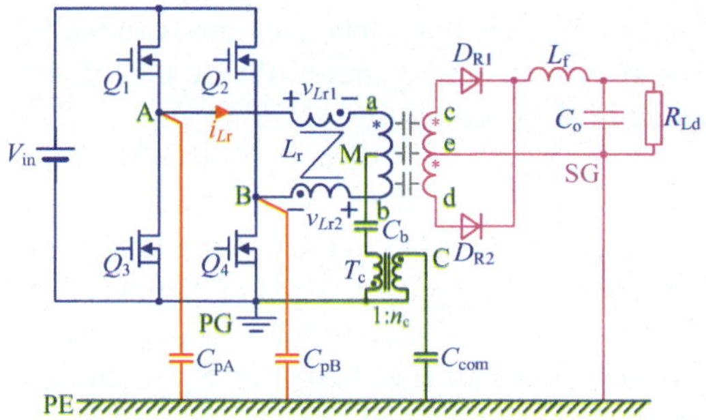

Fig. 8.11 Implementation II for constructing the cancellation circuit

Substituting (8.23) into (8.11), yields

$$v_M = (v_A + v_B)/2 \tag{8.24}$$

As shown in Fig. 8.11, the voltage across point M and the PG is sensed by transformer T_c. C_b is the dc blocking capacitor for removing the DC component in v_M, which is $V_{in}/2$. The secondary winding of T_c outputs a compensation voltage which is out-of-phase with $(v_A + v_B)$. The resultant compensation voltage v_{com} is

$$v_{com} = -n_c\left(\frac{v_A + v_B}{2} - \frac{V_{in}}{2}\right) \tag{8.25}$$

where, n_c is the turns ratio of T_c.

By substituting (8.25) into (8.20), the required compensation capacitance for implementation II is obtained as

$$C_{com} = \frac{C_0 + 2C_p}{n_c} \tag{8.26}$$

Compared with implementation I, only one transformer and one DC blocking capacitor are required in implementation II. Moreover, the size of T_c may be smaller than T_{c1} (T_{c2}), since the volt-second applied to the primary winding of T_c is smaller. Note that, the leakage inductance of T_c should be small enough to make v_{com} be proportional to $(v_A + v_B)$ as desired.

8.4 The Necessity of Combining Two Methods

In this Section, the necessity of combining symmetry circuit and passive cancellation circuit for reducing the CM noise is discussed. The definitions of various solutions are listed in Table 8.1 for the convenience of illustration. Particularly, S_00 is the original case without symmetrical resonant inductor and passive cancellation circuit, and S_11 is the combined solution with both the symmetrical resonant inductor and passive cancellation circuit. Please note that the symmetrical transformer is always used in all solutions.

To compare the CM noise with different solutions, the equivalent noise source (ENS) in Chap. 6 is adopted. Figure 8.12 shows the simplified CM noise model, where, v_S is the ENS and C_S is the total capacitance, as listed in Table 8.2. For S_00, by substituting (8.18) and (8.19) into Fig. 8.5 and simplifying the circuit, v_{S_00} and C_{S_00} can be derived; for S_10, v_{S_10} and C_{S_10} are obtained from the simplified circuit shown in Fig. 8.8b; for S_01, the circuit is similar with Fig. 8.11, while asymmetrical resonant inductor is used and $v_{Lr2} = 0$. The compensation capacitor C_{com} is given as (8.26) to cancel the effect of v_A and v_B; and for S_11, v_{S_11} and C_{S_11} can be calculated according to Fig. 8.9.

According to Fig. 8.12, the CM voltage v_{CM} in the frequency-domain is calculated as

Table 8.1 Definitions of various solutions

Solutions			Definitions
Symmetrical transformer	Asymmetrical resonant inductor	w/o passive cancellation circuit	S_00
		w/ passive cancellation circuit	S_01
	Symmetrical resonant inductor	w/o passive cancellation circuit	S_10
		w/ passive cancellation circuit	S_11

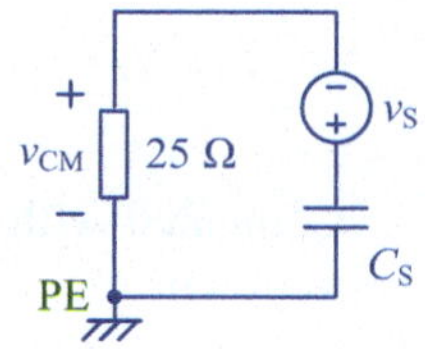

Fig. 8.12 Simplified CM noise model for the PSFB converter

Table 8.2 The ENS and total capacitance for different solutions

S_ij	v_{S_ij}	C_{S_ij}
S_00	$\frac{1}{2}(v_A + v_B) - \frac{1}{2}\frac{C_O}{C_O+2C_p}v_{Lr}$	$C_O + 2C_p$
S_10	$\frac{1}{2}(v_A + v_B)$	$C_O + 2C_p$
S_01	$\frac{n_c}{n_c+1}\frac{2C_p}{C_O+2C_p}\frac{1}{2}v_{Lr}$	$\frac{n_c+1}{n_c}(C_O + 2C_p)$
S_11	0	$\frac{n_c+1}{n_c}(C_O + 2C_p)$

$$\left|\dot{V}_{\mathrm{CM}}(f)\right| = \left|\frac{j2\pi f \cdot 25 \cdot C_{\mathrm{S}}}{j2\pi f \cdot 25 \cdot C_{\mathrm{S}} + 1}\right| \cdot \left|\dot{V}_{\mathrm{S}}(f)\right| \tag{8.27}$$

In Fig. 8.12, the 25-Ω resistor and C_{S} form a high-pass filter with corner frequency f_{c} of $1/(2\pi \cdot 25 \cdot C_{\mathrm{S}})$. Generally, C_{S} is around hundreds of pF, and thus f_{c} is around tens of MHz. Therefore, at the frequencies below tens of MHz, $(2\pi f \cdot 25 \cdot C_{\mathrm{S}}) \ll 1$, and thus (8.27) can be approximated as

$$\left|\dot{V}_{\mathrm{CM}}(f)\right| \approx |2\pi f \cdot 25 \cdot C_{\mathrm{S}}| \cdot \left|\dot{V}_{\mathrm{S}}(f)\right| \tag{8.28}$$

8.4.1 Solution with Only Symmetrical Resonant Inductor

As seen in Table 8.2, for the solutions S_00 and S_10, $C_{\mathrm{S\_00}}$ and $C_{\mathrm{S\_10}}$ are equal, so the difference of the CM noises depends on the spectra difference between $v_{\mathrm{S\_00}}$ and $v_{\mathrm{S\_10}}$, and the spectra of $(v_{\mathrm{A}} + v_{\mathrm{B}})$ and v_{Lr} are studied. The waveforms of $(v_{\mathrm{A}} + v_{\mathrm{B}})$ and v_{Lr} are plotted in Fig. 8.1b. By expanding $(v_{\mathrm{A}} + v_{\mathrm{B}})$ and v_{Lr} within $[0, T_{\mathrm{s}}]$ into Fourier Series, we have

$$\begin{cases} v_{\mathrm{A}} + v_{\mathrm{B}} = V_{\mathrm{in}} + \sum_{k=1,3,5\ldots}^{\infty} \frac{4V_{\mathrm{in}}}{k\pi} \sin\left(\frac{k\varphi}{2}\right) \cos\left(k\omega_{\mathrm{s}}t + \frac{k\varphi}{2}\right) \\ v_{Lr} = \sum_{k=1,3,5\ldots}^{\infty} \frac{4V_{\mathrm{in}}}{k\pi} \sin\left(\frac{k\psi}{2}\right) \cos\left(k\omega_{\mathrm{s}}t - \frac{k\psi}{2}\right) \end{cases} \tag{8.29}$$

where, k is the harmonic order, φ and ψ are defined as

$$\varphi = 2\pi(t_3 - t_2)/T_{\mathrm{s}} \tag{8.30a}$$

$$\psi = 2\pi(t_4 - t_3)/T_{\mathrm{s}} \tag{8.30b}$$

To analyze the influence of v_{Lr} on the CM noise, Fig. 8.13 gives the phasor diagrams of the ENS with solutions S_00 and S_10, where the angle $\theta = k(\varphi + \psi)/2$ according to (8.29). As seen in Fig. 8.13a, when θ is an obtuse angle, eliminating v_{Lr} leads to the reduction of the CM noise, while as seen in Fig. 8.13b, when θ is an acute angle, eliminating v_{Lr} may lead to the increase of CM noise. This implies that the solution with only the symmetry circuit is not effective to reduce the CM noise.

8.4.2 Solution with Only Passive Cancellation Circuit

For the solutions S_00 and S_01, substitution of $C_{\mathrm{S\_00}}$, $C_{\mathrm{S\_01}}$, $v_{\mathrm{S\_00}}$ and $v_{\mathrm{S\_01}}$ in Table II into (8.28), yields

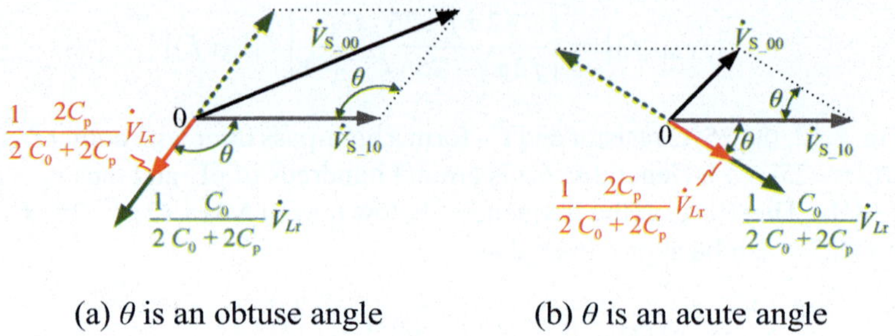

Fig. 8.13 Phasor diagrams of the ENS with solutions S_00, S_10 and S_01

$$\frac{\left|\dot{V}_{\mathrm{CM\_01}}(f)\right|}{\left|\dot{V}_{\mathrm{CM\_00}}(f)\right|}=\frac{\left|\frac{1}{2}\frac{2C_{\mathrm{p}}}{C_0+2C_{\mathrm{p}}}\dot{V}_{Lr}(f)\right|}{\left|\dot{V}_{\mathrm{S\_00}}(f)\right|} \tag{8.31}$$

The phasor of $\frac{1}{2}\frac{2C_{\mathrm{p}}}{C_0+2C_{\mathrm{p}}}\dot{V}_{Lr}$ is also given in Fig. 8.13. Note that a planar transformer has large inter-winding capacitance, so C_0 is usually much larger than C_{p}, and the magnitude of phasor $\frac{1}{2}\frac{2C_{\mathrm{p}}}{C_0+2C_{\mathrm{p}}}\dot{V}_{Lr}$ will be relatively small than that of $\frac{1}{2}\frac{C_0}{C_0+2C_{\mathrm{p}}}\dot{V}_{Lr}$. So, with only the passive cancellation circuit, the CM noise can be reduced.

According to the above discussion, it can be concluded that the solution with only the symmetry circuit will not be effective to reduce the CM noise, while the solution with only passive cancellation circuit can reduce the CM noise. Furthermore, when the symmetry circuit and the passive cancellation circuit are combined, the CM noise of the PSFB can be effectively reduced.

8.5 Experimental Verification

8.5.1 Descriptions of the Prototype

In order to verify the proposed methods for cancelling the CM noise, a PSFB converter prototype is fabricated and tested in the lab, as shown in Fig. 8.14. The main circuit parameters are listed in Table 8.3. A 2.5-μF DC blocking capacitor C_{block} is connected in series with the primary winding to avoid the saturation of transformer, which can be treated as short circuit owing to its far smaller dv/dt than v_{A}, v_{B} and v_{Lr}.

The power transformer uses PQI 50/33 ferrite core. The transformer winding structure has been shown in Fig. 8.7b, and the corresponding parameters are listed in Table 8.4. The primary windings Pri1 and Pri2 are both tightly wound layers with the same turns. The same type of insulation materials with the same thickness are inserted between Pri1 and Sec1-1 and between Pri2 and Sec2-2, and thus C_{01} and C_{02} are nearly the same. As seen in Fig. 8.7b, the power transformer is a symmetrical transformer, and it is used throughout the experiments. Besides, the parasitic

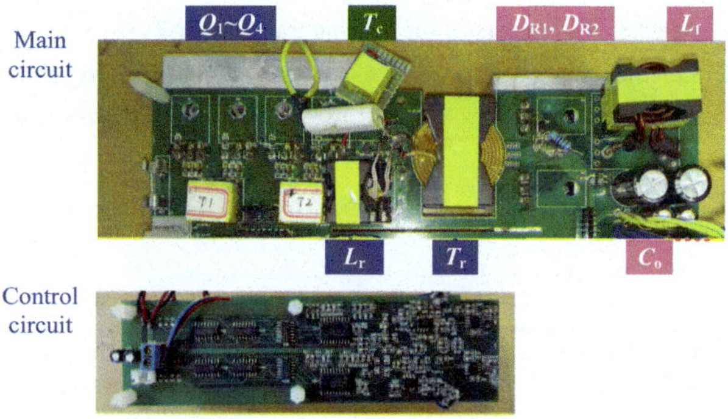

Fig. 8.14 Photograph of the prototype

Table 8.3 Main circuit parameters of the PSFB converter

Input voltage V_{in}	380 V	Resonant inductance L_r	30 μH
Output voltage V_o	48 V	Input capacitance C_{in}	2.5 μF
Output power P_o	1 kW	Output filter inductor L_f	13 μH
Switching frequency f_s	100 kHz	Output filter capacitor C_o	2400 μF
Transformer turns ratio n	1/6	DC blocking capacitor C_{block}	2.5 μF

capacitances between two phase-leg midpoints and the heat sink are nearly the same. The measured capacitances are: $C_{01} = C_{02} = 50$ pF; $C_p = C_{pA} = C_{pB} = 36$ pF (the parasitic capacitance from the phase-leg midpoints to the heat sink is 24 pF, and the inter-winding capacitance of the driving transformer which connects the main switches and the SG is 12 pF).

PQ 26/20 ferrite core and 0.1 mm × 140 Litz wire are adopted for both the asymmetrical and symmetrical resonant inductor. The asymmetrical resonant inductor uses one winding with 8 turns, while the symmetrical resonant inductor uses two windings, and each winding has 4 turns. The implementation II shown in Fig. 8.11 is used for realizing the passive cancellation circuit. For the transformer T_c, single strand enameled wire is adopted for the windings. The primary and secondary windings are wound side by side in each layer to reduce the leakage inductance. The parameters of T_c are listed in Table 8.5. The DC blocking capacitor C_b is 2 μF, and the compensation capacitance C_{com} is 176 pF.

Table 8.4 Parameters of the power transformer

	Wire	Winding dimension	Turns
Primary winding ab	Litz wire	0.1 mm × 140	Pri1: 6
			Pri2: 6
Secondary winding cd	PCB winding	4 oz, 9.75 mm	4

Table 8.5 Parameters of the sensing transformer T_c

Implementation		Measured parameters	
Core	EP 13	Magnetizing inductance	3.9 mH
Winding dimension	0.25 mm	Leakage inductance	0.38 μH
Turns ratio	52:52	Structural capacitance	198 pF

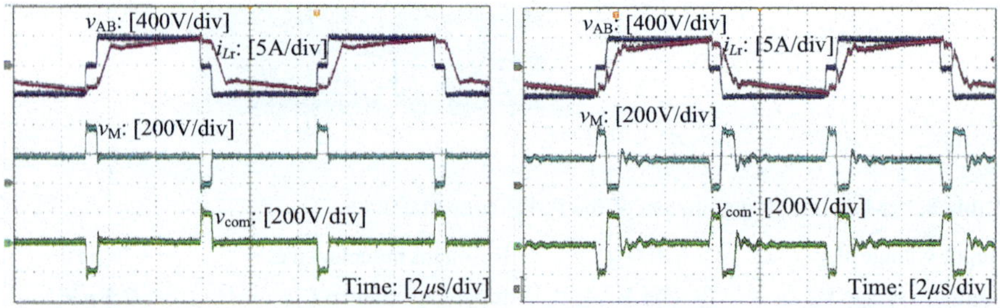

(a) With symmetrical resonant inductor (b) With asymmetrical resonant inductor

Fig. 8.15 Voltage waveforms of nodes M and C with respect to the PG

8.5.2 Measurement and Comparison of the CM Noise

Figure 8.15 shows the measured waveforms of the PSFB converter at full load with the passive cancellation circuit. It can be observed that v_M and v_{com} are out-of-phase. In Fig. 8.15a, v_M equals to $(v_A + v_B)/2$ when using the symmetrical resonant inductor. In Fig. 8.15b, v_M equals to $(v_A + v_B - v_{Lr})/2$ when adopting the asymmetrical resonant inductor.

Figure 8.16 shows the key experimental waveforms of the PSFB converter with the solutions listed in Table 8.1. The CM noise voltage v_{CM} is measured with an oscilloscope which uses 50-Ω input impedances.

As seen in Fig. 8.16, an impulse is presented in v_{CM} when the voltages of the midpoints A and B have step change at t_1 and t_2, respectively. When the resonant inductor voltage has step change at t_3, an impulse occurs in v_{CM}, as shown in Fig. 8.16a, c, while no impulse is observed in v_{CM}, as shown in Fig. 8.16b, d. This implies that when the symmetrical resonant inductor and symmetrical transformer are both adopted, the influence of the resonant inductor voltage on CM noise is eliminated.

Figure 8.17 shows the simulated and measured spectra for the CM noise of the PSFB converter at full load with different solutions. Here, the simulated spectra for the CM noise is represented by the CM noise envelope, which is obtained by connecting every two adjacent noise peaks in the simulated spectra with a straight line, for the purpose of avoiding the overlap between the simulated and measured results and offering a clear comparison. As seen, from 150 kHz to 1 MHz, the measured and simulated results are well matched; while at higher frequencies,

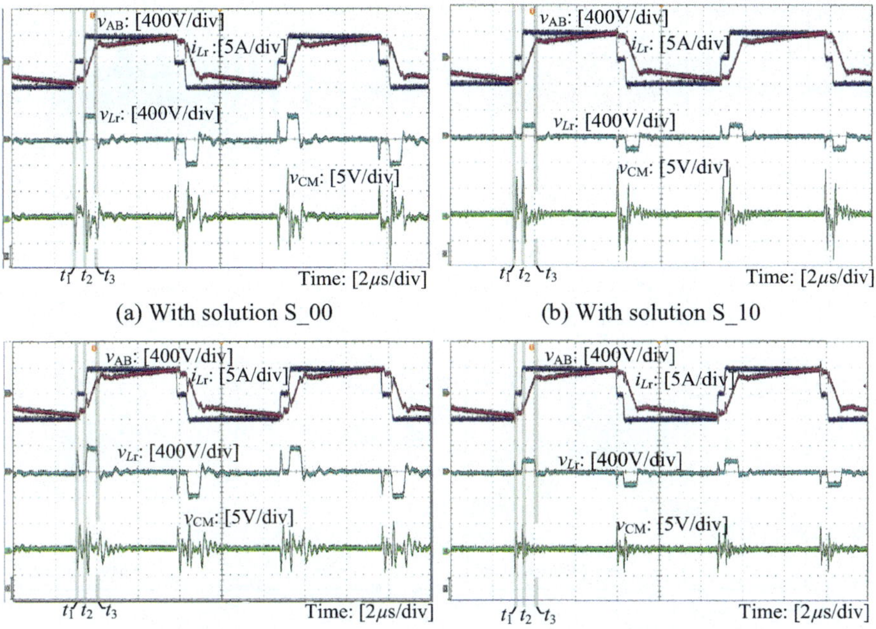

(a) With solution S_00

(b) With solution S_10

(c) With solution S_01

(d) With solution S_11

Fig. 8.16 Key waveforms of PSFB converter and the measured CM noise voltage

there are differences between these spectra, due to some parasitic elements that are neglected in the model. It should be noted that the simulated result is lower than the measured spectra in Fig. 8.17d, because the related parasitic capacitances are not exactly matched in the experiment, degrading the practical CM noise reduction.

Figure 8.18 shows the comparison of the measured CM noises with different solutions. In Fig. 8.18a, the CM noise of the solution S_10, which only uses the symmetrical resonant inductor, is given and compared with the original CM noise (S_ 00). As seen, the spectra difference varies with the frequency. Although the resonant inductor voltage is eliminated by the symmetrical resonant inductor, the CM noise is lower than the original one at some frequencies, and it is even higher than the original one at other frequencies. This means the solution with only the symmetrical resonant inductor can not effectively reduce the CM noise, which is in agreement with the analysis in Sect. 8.4.

In Fig. 8.18b, the CM noise of the solution S_01, which only adopts the passive cancellation circuit, is given together with the original one. As seen, with only the passive cancellation circuit, the CM noise is reduced by several dB at the frequencies below 1 MHz, which is in agreement with the analysis in Sect. 8.4.

Figure 8.18c illustrates the measured CM noise of the solution S_11, which is the combined solution with both the symmetrical resonant inductor and passive cancellation circuit. Meanwhile, the original CM noise is also given for comparison. As

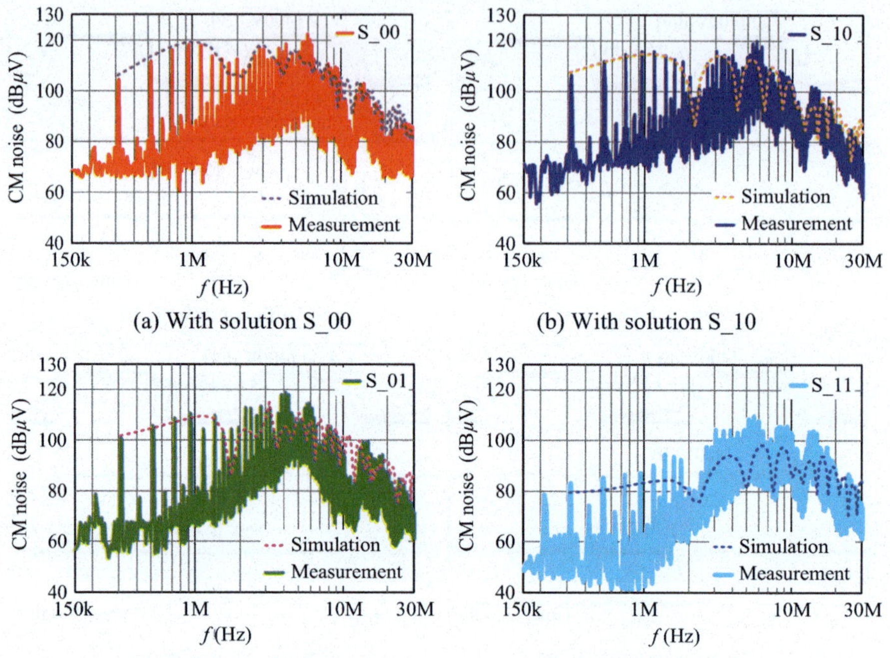

(a) With solution S_00 (b) With solution S_10

(c) With solution (d) With solution S_11

Fig. 8.17 Comparison of the simulated and measured CM noise of PSFB converter

seen, the CM noise is significantly reduced by the combined method. From 150 kHz to 3 MHz, the CM noise is reduced by 20 dB; from 3 to 10 MHz, the amount of noise reduction decreases with the increasing of frequency; from 10 to 30 MHz, the noise reduction is not so effective. This is due to the increased winding resistance of T_c at high frequency, which limits the compensation current and reduces the effectiveness of cancellation.

Figure 8.18d gives the measured CM noises of the solutions S_10, S_01 and S_11. As seen, the combined solution achieves the best CM noise reduction. This verifies the necessity of combining the symmetry circuit and passive cancellation circuit in Sect. 8.4.

8.6 Summary

In this chapter, the CM noise model for the PSFB converter is derived, where, the noise voltage sources include the voltages between two phase-leg midpoints and the PG, and the resonant inductor voltage. To reduce the CM noise, the symmetrical resonant inductor and symmetrical transformer is proposed, which eliminates the effect of resonant inductor voltage on the CM noise. Then, a passive cancellation

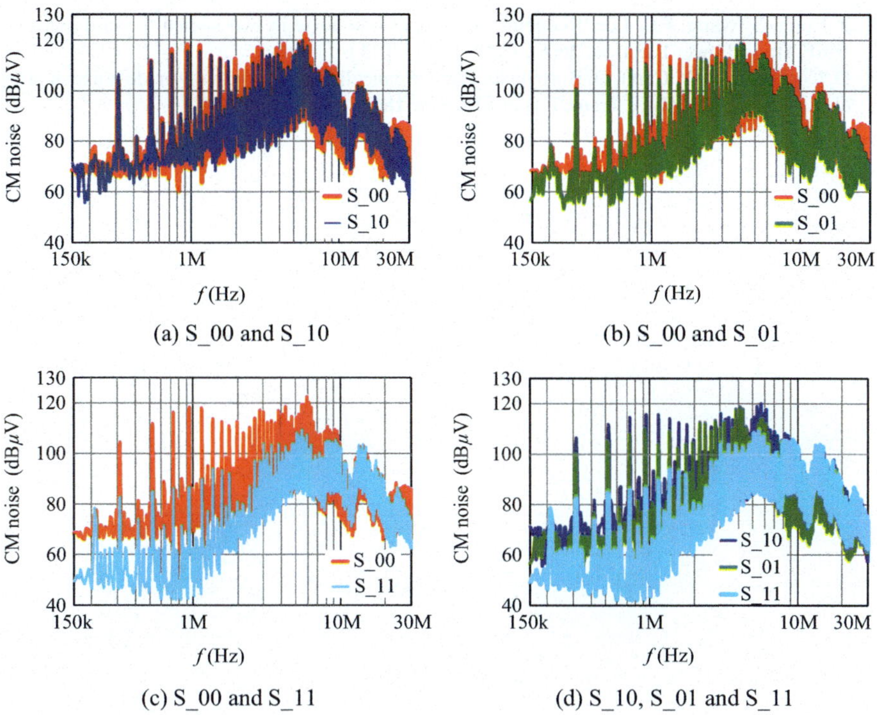

Fig. 8.18 The comparison of measured CM noises of the PSFB converter

circuit consisting of a compensation voltage source and a compensation capacitor is introduced, which eliminates the effect of the voltages between the two phase-leg midpoints and the PG. A 1-kW prototype of the PSFB converter is fabricated in the lab, and the CM noise reduction by the symmetrical circuit, passive cancellation circuit and the combined solution are examined, respectively. The measured CM noise demonstrates that both the symmetrical circuit and passive cancellation are necessary for achieving significant CM noise reduction.

References

1. Y. Chu, S. Wang, A generalized common mode current cancellation approach for power converters. IEEE Trans. Industr. Electron. **62**(7), 4130–4140 (2015)
2. H. Zhang, S. Wang, Y. Li, Q. Wang, D. Fu, Two-capacitor transformer winding capacitance models for common-mode EMI noise analysis in isolated dc–dc converters. IEEE Trans. Power Electron. **32**(11), 8458–8469 (2017)
3. J. He, W. Chen, and J. Jiang, Model and analysis on common-mode conducted EMI of a switched mode power supply, in Proceedings of the Chinese Society for Electrical Engineering (CSEE), vol. 25, no. 8, pp. 50–55, 2005 (in Chinese)

4. P. Kong, S. Wang, F.C. Lee, Reducing common-mode noise in two-switch forward converter. IEEE Trans. Power Electron. **26**(5), 1522–1533 (2011)
5. D. Cochrane, D.-Y. Chen, D. Boroyevic, Passive cancellation of common-mode noise in power electronic circuits. IEEE Trans. Power Electron. **18**(3), 756–763 (2003)

Chapter 9
Reducing the Common-Mode (CM) Noise in DC-DC Converters by the CM Voltage Cancellation Method

9.1 The Common-Mode Voltage Cancellation Method

In Chap. 6, the common-mode (CM) noise equivalent circuit of DC-DC converters is derived, as shown in Fig. 9.1a, where v_{ENS} is the equivalent noise source (ENS), C_{sum} is the total parasitic capacitance, and the 25-Ω resistor is the equivalent impedance for the CM noise at the line impedance stabilization network (LISN) side [1]. For the passive cancellation method [2], the cancellation branch consisting of a compensation voltage v_{com} and a compensation capacitor C_{com} is in parallel with the converter, as shown in Fig. 9.1b, which cancels the current that generates the CM noise in the converter.

According to the duality, a compensation voltage v_{com} can be inserted and connected in series with v_{ENS}, as shown in Fig. 9.1c. If $v_{\mathrm{com}} = v_{\mathrm{ENS}}$, then v_{CM} is zero, and the CM noise is eliminated. This method reduces the CM noise by cancelling the CM voltage of DC-DC converters, and it is identified as the CM voltage cancellation (CMVC) method. The CMVC method is similar with the CM transformer approach that eliminates the leakage current in motor drive systems [3].

For implementing the CMVC method, the key is how to obtain the CM voltage. In a DC-DC converter, v_{ENS} is generally expressed as [1]

$$v_{\mathrm{ENS}} = k_{\mathrm{C}} v_{\mathrm{Q}} \tag{9.1}$$

where, k_{C} is the ratio related with the parasitic capacitances that affects the CM noise, and v_{Q} is the noise voltage source, which is the drain-to-source voltage of power switch.

In general, the voltage v_{mag} across the magnetic component have the same AC component as v_{Q} [2], i.e.,

$$v_{\mathrm{Q}} = v_{\mathrm{mag}} \tag{9.2}$$

© The Author(s), under exclusive license to Springer Nature Singapore Pte Ltd. 2024

X. Ruan et al., *Conducted Electromagnetic Interference in Power Converters: Modeling, Prediction and Reduction*, CPSS Power Electronics Series,
https://doi.org/10.1007/978-981-99-9295-9_9

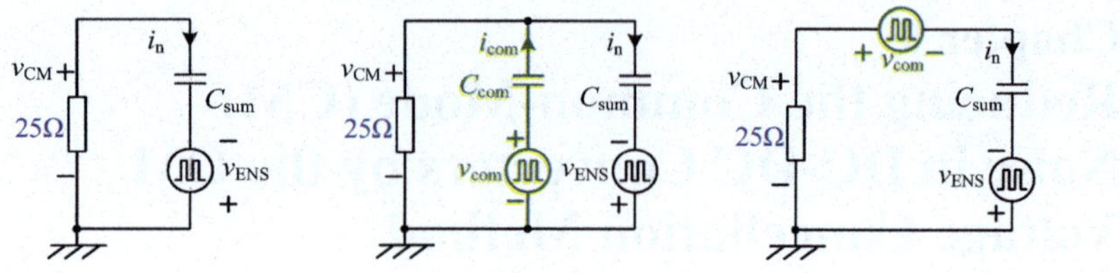

(a) CM noise equivalent circuit (b) CM current cancellation (c) CM voltage cancellation

Fig. 9.1 Principles of the CM noise cancellation method

As seen from Fig. 9.1c, for achieving the CMVC, the compensation voltage source v_{com} is required to be equal to v_{ENS}. Referring to (9.1) and (9.2), we have

$$v_{com} = k_C v_{mag} \tag{9.3}$$

According to (9.3), v_{com} can be obtained by adding auxiliary windings into the magnetic component (it could be an inductor or a transformer), as shown in Fig. 9.2a.

Based on Fig. 9.2a, the compensation voltage kv_{mag} are added into both the input power cords in a DC-DC converter, as shown in Fig. 9.2b, where, the windings AB and CD are the CMVC windings, and their turns N_{AB} and N_{CD} are both equal to $k_C N_{PQ}$, and N_{PQ} is the turns of the winding with voltage of v_{mag}. By neglecting the leakage inductance between AB and CD, v_{AB} is equal to v_{CD}. Obviously, the added CMVC windings do not affect the normal operation of the power converter. In addition, by placing the input filtering capacitor C_{in} between the CMVC windings and the power converter, the current flowing through the CMVC winding is almost a DC current, minimizing the size and losses of the CMVC windings.

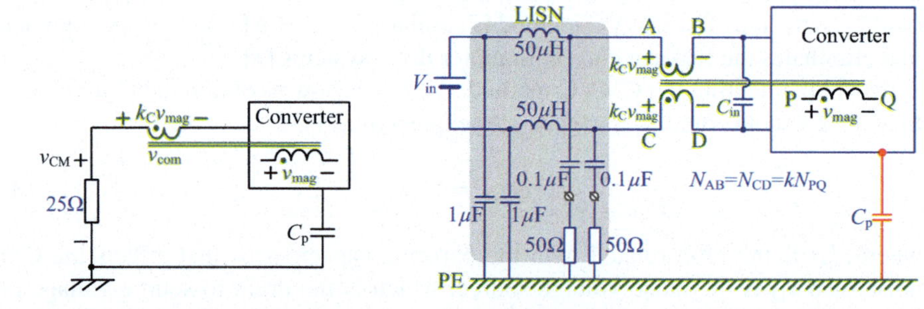

(a) Principle of the CMVC method (b) Realization of the CMVC method

Fig. 9.2 Principle and realization of the CMVC method

9.2 Application of the CMVC Method in Non-isolated DC-DC Converters

In this Section, the application of the CMVC method in buck converter is first studied. Figure 9.3a shows the buck converter with the CM parasitic capacitances, where, C_{p1} is the parasitic capacitance between the switching node and the PE, C_{p2}, C_{p3} and C_{p4} are the parasitic capacitances between the positive terminal of the output voltage, power ground, drain of power switch Q and the PE, respectively.

The equivalent CM noise circuit is derived with the substitution theorem. Note that, when applying the substitution theorem, voltage loops or current nodes should be avoided. Firstly, the freewheeling diode D_{FW} is replaced by the voltage source v_{DFW} with its own waveform, owing to the high dv/dt node M that causes the CM noise. Then, the output capacitor is treated as short circuit because its dv/dt is far lower than that of v_{DFW}. To avoid voltage loops, power switch Q and the output filter inductor L_f are substituted by current sources with their own waveform, respectively. Finally, by considering the 25-Ω equivalent resistor from the LISN side, the CM noise equivalent circuit is derived, as shown in Fig. 9.3b.

To derive the ENS, the equivalent circuit shown in Fig. 9.3b is further simplified with the superposition theorem. Firstly, the voltage source v_{DFW} is shorted to solely analyze the effect of the current sources. Because all the current sources are essentially shorted, they have no effect on the CM noise. Secondly, by opening all current sources, the excitation of v_{DFW} is individually applied to the circuit. By adopting the Thevenin's equivalent circuit method, the simplified CM noise circuit is derived, as shown in Fig. 9.3c, where

$$C_{sum} = C_{p1} + C_{p2} + C_{p3} + C_{p4} \tag{9.4}$$

$$k_{buck} = \frac{C_{p1}}{C_{p1} + C_{p2} + C_{p3} + C_{p4}} = \frac{C_{p1}}{C_{sum}} \tag{9.5}$$

Referring to Fig. 9.3a, v_{DFW} and v_{Lf} have the same AC components, so the ENS can be also expressed as $k_{buck}v_{Lf}$. Referring to Figs. 9.2b and 9.4 shows the buck

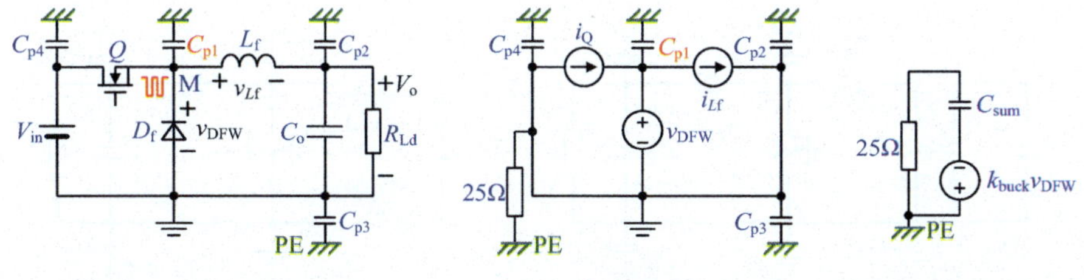

(a) Buck converter (b) CM noise model (c) Simplified circuit

Fig. 9.3 Buck converter and its CM noise model

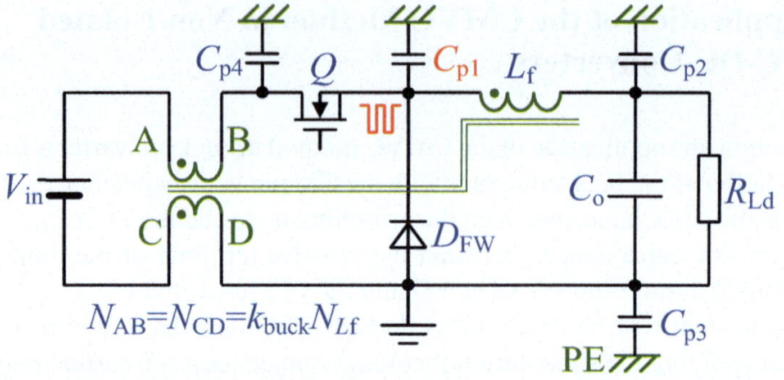

Fig. 9.4 Buck converter with the CMVC method

converter adopting the CMVC method. The turns of the CMVC windings AB and CD are both equal to $k_{buck}N_{Lf}$, where N_{Lf} is the turns of the winding for L_f.

Following the procedure for applying the CMVC method in buck converter, Fig. 9.5a shows the boost converter with the CMVC method. The ENS of boost converter is $k_{bst}v_Q$, where $k_{bst} = C_{p1}/(C_{p1} + C_{p2} + C_{p3})$, v_Q is the drain-source voltage of Q. The turns of the CMVC winding AB and CD are equal to $k_{bst}N_{Lb}$, where N_{Lb} is the turns of L_b.

As seen from Fig. 9.5a, the right-side node of L_b and node A of winding AB are dotted terminals, so the windings of L_b and AB can be combined into one winding with total turns of $(1–k_{bst})N_{Lb}$, as shown in Fig. 9.5b. This circuit has been proposed in [4] by adopting the CM noise balancing. The upside and downside coupled inductors are essentially in series, and their total turns is equal to the turns of the original inductor L_b, implying that this method does not increase the size of boost inductor.

Figure 9.6 gives the applications of CMVC method in other basic non-isolated DC-DC converters. For the buck-boost converter, the turns ratio k_{bb} of the CMVC winding and the winding of L_f is $C_{p1}/(C_p + C_{p2} + C_{p3} + C_{p4})$; for the Cuk and SEPIC converter, the corresponding turns ratio k_{Cuk} and k_{SEPIC} are equal to $(C_{p1} + C_{p2})/(C_{p1} + C_{p2} + C_{p3} + C_{p4})$; and for Zeta converter, the turns ratio k_{Zeta} is equal to $(C_{p1} + C_{p2})/(C_{p1} + C_{p2} + C_{p3} + C_{p4} + C_{p5})$.

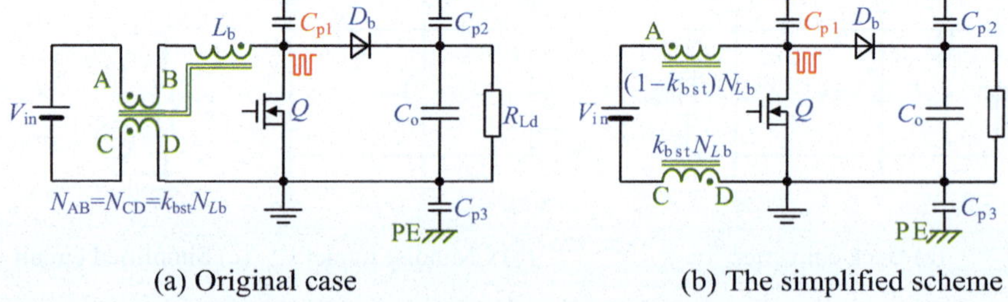

Fig. 9.5 Boost converter with the CMVC method

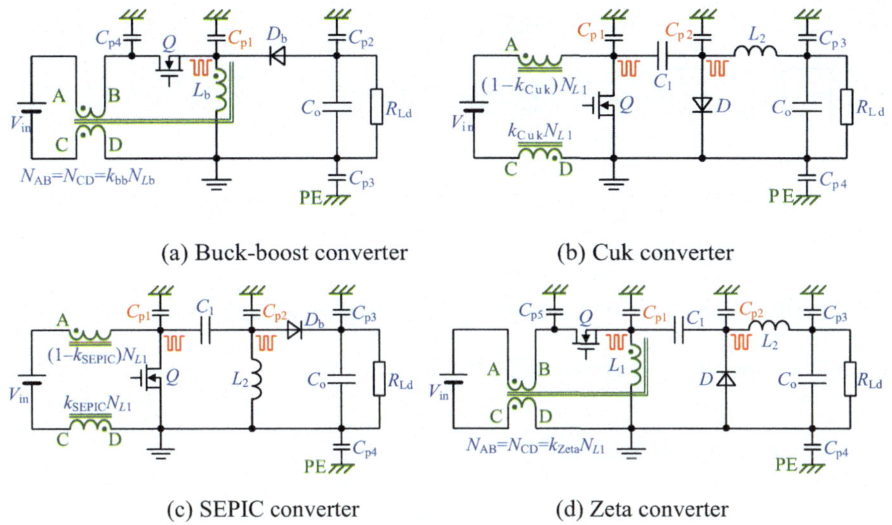

Fig. 9.6 Other basic non-isolated DC-DC converters with the CMVC method

According to the above discussions, it is concluded that for non-isolated DC-DC converters, without input side inductor (buck, buck-boost and Zeta), the CMVC windings at the input side is required; while for the ones with input side inductor (boost, Cuk and SEPIC), the CMVC method can be achieved by splitting the input side inductor into coupled inductors.

9.3 Application of the CMVC Method in Isolated DC-DC Converters

In non-isolated DC-DC converters, the CMVC windings are coupled with inductor winding to nullify the CM voltage. While in isolated DC-DC converters, the CMVC windings are considered to be coupled with transformer windings.

The CM noise circuit of isolated DC-DC converters can be derived by replacing the inter-winding capacitance of the transformer with two-capacitor model [5] and substituting the switches with voltage or current sources. Figure 9.7a shows the flyback converter adopting the CMVC method, where, PG and SG represent the primary and secondary ground, respectively, and v_Q denotes the drain-to-source voltage of power switch Q. The two-capacitor model for the transformer is given in Fig. 9.7b [5]. By substituting Q with the voltage source and the rectifier diode D_o with the current source, and simplifying the resultant circuit, the ENS is given by

$$v_{ENS\_FLY} = k_{FLY} v_Q \tag{9.6}$$

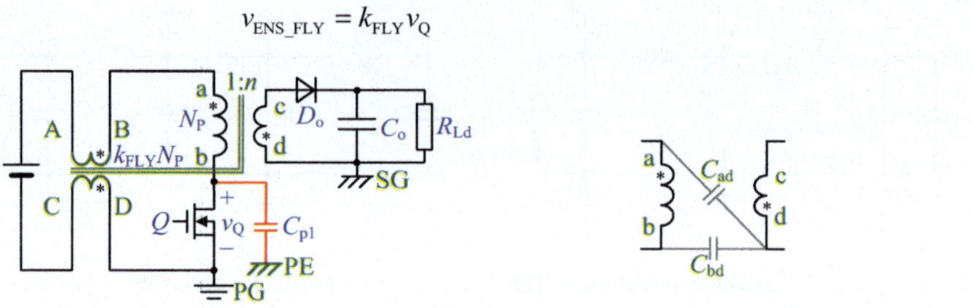

(a) Circuit with the CMVC method (b) Corresponding two-capacitor model

Fig. 9.7 The flyback converter with the CMVC method

where, k_{FLY} is expressed as

$$k_{FLY} = \frac{C_{bd} + C_{p1}}{C_{ad} + C_{bd} + C_{p1}} \tag{9.7}$$

Since the voltage v_{ab} across the primary winding ab and v_Q have the same AC component, the turns of the CMVC winding AB and CD should be both equal to $k_{FLY}N_P$ to cancel the CM voltage, as shown in Fig. 9.7a. Where, N_P is the turns of the primary winding.

Since the CMVC windings are located at the primary-side circuit in isolated DC-DC converters, the electric field coupling between the CMVC and secondary windings should be prevented, so the primary winding is intentionally placed between them. Such arrangement has the merit that the magnetic coupling of the primary and secondary windings is not affected. Figure 9.8 shows an example of the transformer with the above winding arrangement, where Pri1 ~ Pri4 are the primary-winding layers connected in series, and Sec is the secondary-winding layer, and CM1 and CM2 are the CMVC windings.

Fig. 9.8 The transformer winding structure of flyback converter with the CMVC method

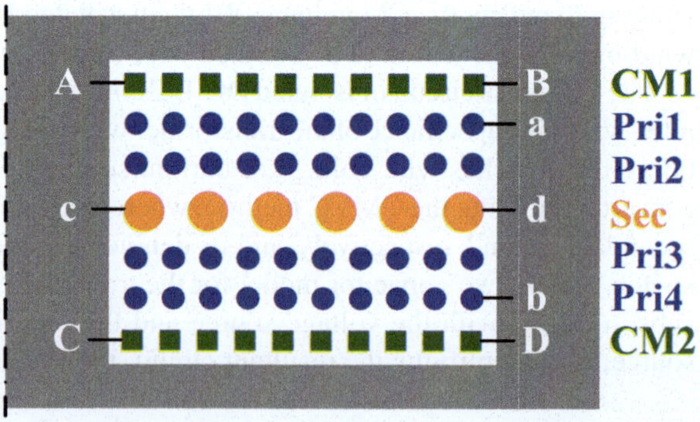

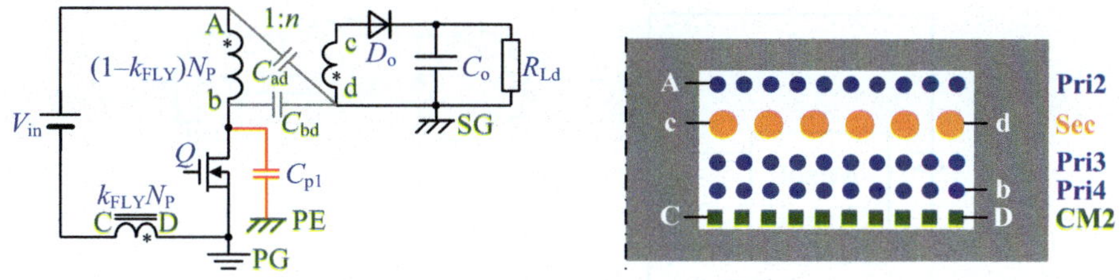

(a) Flyback converter (b) Transformer winding structure

Fig. 9.9 The simplified flyback converter with the CMVC method

Referring to Fig. 9.7a, the CMVC winding AB and the transformer primary winding ab are connected in series at the dotted terminals. So, they can be combined, as shown in Fig. 9.9a, where the turns of the primary winding is changed to $(1 - k_{\text{FLY}})N_{\text{p}}$. As a result, for the transformer shown in Fig. 9.8, the CM1 and Pri1 winding layer can be removed (the turns of CM1 and Pri1 are assumed to be equal for better illustration), as shown in Fig. 9.9b. The Pri2, Pri3, Pri4 and CM2 are essentially connected in series, which has equal turns as the original winding ab. Therefore, the transformer size would not be increased when adopting the CMVC method. The circuit configuration in Fig. 9.9a has been proposed in [6], which is derived by combining symmetry circuit and winding cancellation method to reduce the CM noise.

The applications of the CMVC method in other isolated DC-DC converters are given in Fig. 9.10. In terms of the forward, push–pull, and half-bridge (HB) converters, the turns ratios of the CMVC windings and the primary winding are expressed as

$$k_{\text{FWD}} = \frac{C_{\text{bd}} + C_{\text{p1}}}{C_{\text{ad}} + C_{\text{bd}} + C_{\text{p1}}} \tag{9.8}$$

$$k_{\text{PP}} = \frac{C_{\text{ae}} - C_{\text{be}} + C_{\text{p1}} - C_{\text{p2}}}{C_{\text{ae}} + C_{\text{be}} + C_{\text{p1}} + C_{\text{p2}}} \tag{9.9}$$

$$k_{\text{HB}} = \frac{C_{\text{ae}} + C_{\text{p2}}}{C_{\text{ae}} + C_{\text{be}} + C_{\text{p1}} + C_{\text{p2}}} \tag{9.10}$$

For the the HB *LLC* resonant converter, as shown in Fig. 9.10d, C_{p1} and C_{p2} are the parasitic capacitances between the drain of Q_1, Q_2 and the PE through the heatsink. As the switching frequency is close to the resonant frequency, the voltage across the L_{r} and C_{r} branch can be neglected since its impedance is far less than that of the magnetizing inductance. Therefore, the CM noise model of the HB *LLC* resonant converter and the HB converter is similar, and the turns ratio $k_{\text{HB\_LLC}}$ is given by

$$k_{\text{HB\_LLC}} = \frac{C_{\text{ae}} + C_{\text{p2}}}{C_{\text{ae}} + C_{\text{be}} + C_{\text{p1}} + C_{\text{p2}}} \tag{9.11}$$

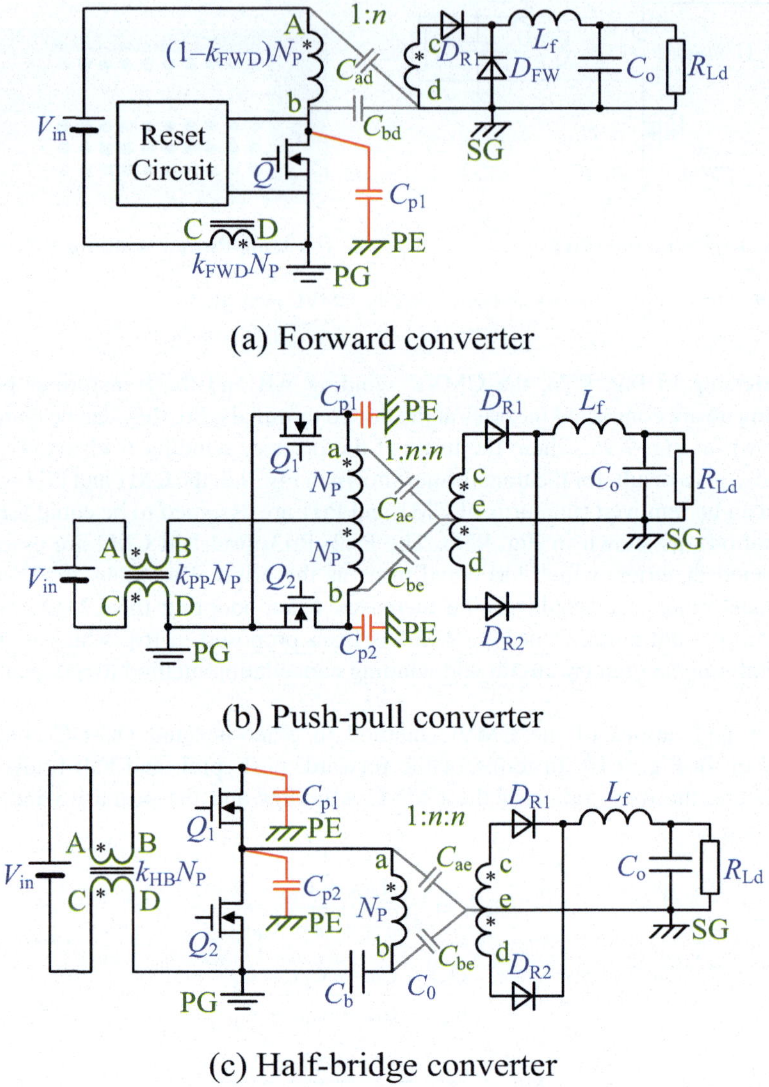

Fig. 9.10 The CMVC method for basic isolated DC-DC converters

Figure 9.10e shows the full-bridge converter, where C_{p1} ~ C_{D4} are the parasitic capacitance between the drain of Q_1 ~ Q_4 to the PE through the heatsink. The ENS is given by [6]

$$v_{\text{ENS\_FB}} = \frac{(C_{\text{ae}} + C_{\text{p2}})v_{\text{A}} + (C_{\text{be}} + C_{\text{p4}})v_{\text{B}} - C_{\text{ae}}v_{Lr}}{C_{\text{ae}} + C_{\text{be}} + C_{\text{p1}} + C_{\text{p2}} + C_{\text{p3}} + C_{\text{p4}}} \tag{9.12}$$

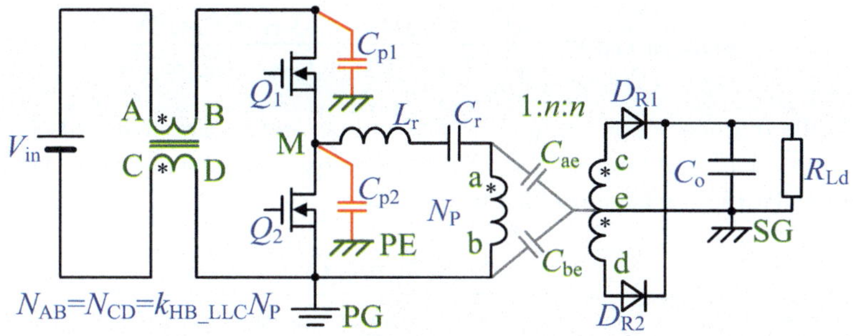

(d) Half-bridge *LLC* resonant converter

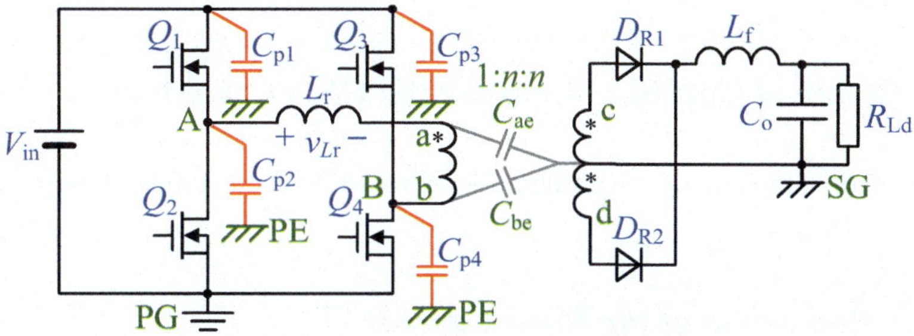

(e) Full-bridge converter

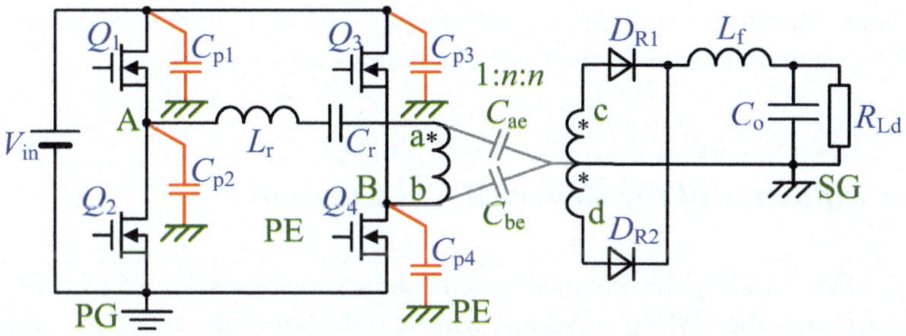

(f) Full-bridge *LLC* resonant converter

Fig. 9.10 (continued)

As seen from Fig. 9.10e, the primary voltage of the transformer, v_{ab}, equals to $v_A - v_B - v_{Lr}$. By comparing (9.12) and v_{ab}, it can be found that the ENS and v_{ab}, are not proportional. Thus, the CMVC method can not be applied in the full-bridge converter.

Figure 9.10f shows the full-bridge *LLC* resonant converter. When the switching frequency is close to the resonant frequency, the voltage across the resonant branch (L_r and C_r in series) can be neglected, and the ENS is given by

$$v_{\text{ENS\_FB\_LLC}} = \frac{(C_{\text{ae}} + C_{\text{p2}})v_{\text{A}} + (C_{\text{be}} + C_{\text{p4}})v_{\text{B}}}{C_{\text{ae}} + C_{\text{be}} + C_{\text{p1}} + C_{\text{p2}} + C_{\text{p3}} + C_{\text{p4}}} \quad (9.13)$$

With the frequency control, Q_1 and Q_4, and Q_2 and Q_3 are switched simultaneously, respectively, and $v_{\text{A}} = -v_{\text{B}}$. Thus, v_{ab} is approximately equal to $2v_{\text{A}}$. Then, the ENS in (9.13) can be simplified as

$$v_{\text{ENS\_FB\_LLC}} = \frac{(C_{\text{ae}} - C_{\text{be}}) + (C_{\text{p2}} - C_{\text{p4}})}{C_{\text{ae}} + C_{\text{be}} + C_{\text{p1}} + C_{\text{p2}} + C_{\text{p3}} + C_{\text{p4}}} v_{\text{A}} \quad (9.14)$$

Therefore, the CMVC method can be applied in the FB *LLC* resonant converter.

9.4 Practical Considerations for the CMVC Method

In this section, the practical considerations for the CMVC method will be presented.

9.4.1 Restriction of the Input Current

Since the input current fully passes through the CMVC windings, the CMVC method is recommended to be applied in the converter with low input current, where the copper loss and dimension of the CMVC windings are acceptable.

9.4.2 Influence of the Reflected CM Currents

As the CMVC windings are coupled with the inductor or transformer, the CM current flowing through the CMVC windings will be reflected to the inductor/transformer side. In general, the amplitude of the CM current is very small compared with that of the current flowing through the inductor/transformer winding, and the CM current is further attenuated by the CMVC method. As a result, the effect of reflected CM current can be neglected.

9.4.3 Balancing Capacitor for Achieving the CMVC

Practically, the parasitic capacitances in the circuit take arbitrary value within reasonable range, so the turns ratio of the CMVC windings and their coupled winding will not be exactly equal to the capacitance ratio under all conditions. To solve this issue, balancing capacitor can be introduced and connected in parallel with some parasitic capacitance, in order to adjust the capacitance ratio for achieving the CMVC.

Taking the CMVC method in buck converter in Fig. 9.4 as an example, by adding a balancing capacitor C_B in parallel with C_{p3}, the condition of achieving CMVC is

$$k_{buck} = \frac{N_{AB}}{N_{Lf}} = \frac{C_{p1}}{C_{p1} + C_{p2} + (C_{p3} + C_B) + C_{p4}} \tag{9.15}$$

By adjusting the capacitance of C_B, the capacitance ratio at the right side will be modified so that its value is equal to the turns ratio at the left side. In fact, C_B can be in parallel with any parasitic capacitance, as long as the capacitance ratio and the turns ratio are equal.

In some converters, the CMVC can be achieved without adding a balancing capacitor. For push–pull converter in Fig. 9.10b, when the insulation between Q_1, Q_2 and the heatsink has the same thickness, $C_{p1} = C_{p2}$ can be assumed. Meanwhile, by adopting symmetry transformer [7], $C_{ae} = C_{be}$ can be obtained. According to (9.9), $k_{PP} = 0$, which means no CM noise. This is also true for the FB *LLC* resonant converter: when $C_{ae} = C_{be}$ and $C_{p1} = C_{p2}$, the CM noise can be eliminated according to (9.14).

For the HB converter and HB *LLC* resonant converter, by using the insulation with the same thickness between Q_1, Q_2 and the heatsink, and adopting symmetry transformer, $C_{p1} = C_{p2}$ and $C_{ae} = C_{be}$ can be realized. According to (9.10) and (9.11), k_{HB} and k_{HB_LLC} are both equal to 0.5.

Taking the transformer winding structure shown in Fig. 9.11 as an example, the condition for achieving $C_{ae} = C_{be}$ will be derived. In Fig. 9.11, the primary winding ab is contributed by two planar windings Pri1 and Pri2 connected in parallel, the secondary winding cd is implemented by printed circuit board (PCB) winding with the center-tap terminal e. The CMVC windings, AB and CD, are placed at the top side. The parasitic capacitance between Pri1 and Sec1-1, Pri2 and Sec2-2 are denoted by C_{01} and C_{02}, respectively. The displacement current flowing through Pri1 and Sec1-1, Pri2 and Sec2-2 are expressed as

$$\begin{cases} i_{dis1} = C_{01}\frac{d}{dt}\left[\lambda_1 v_a + (1-\lambda_1)v_b - \left(\frac{7}{8}v_c + \frac{1}{8}v_d\right)\right] \\ i_{dis2} = C_{02}\frac{d}{dt}\left[(1-\lambda_1)v_a + \lambda_1 v_b - \left(\frac{1}{8}v_c + \frac{7}{8}v_d\right)\right] \end{cases} \tag{9.16}$$

where, λ_1 is given by

$$\lambda_1 = \frac{\frac{2}{3}r_1 + \frac{1}{3}r_2 + \frac{r_2 - r_1}{6N_P^2}}{r_1 + r_2} \tag{9.17}$$

where, r_1 and r_2 are the inner and outer radius of the primary winding, N_P is the turns number of the primary winding [8].

Because one insulation with the same thickness is used between Pri1 and Sec1-1, as well as Pri2 and Sec2-2, $C_{01} = C_{02} = 0.5C_0$ is ensured. According to (9.16), and considering $C_{01} = C_{02} = 0.5C_0$, we have

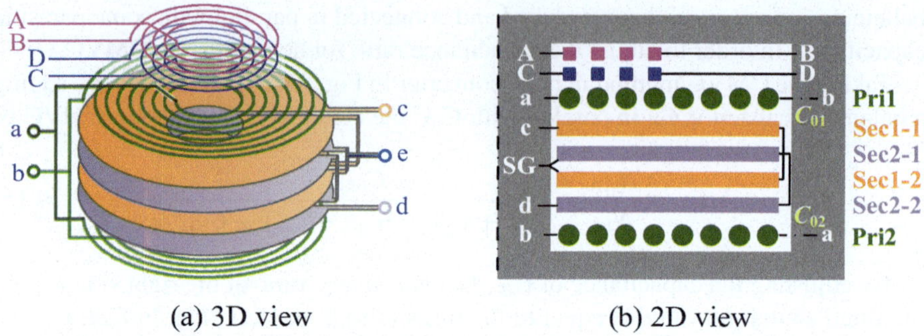

Fig. 9.11 Winding structure of the transformer

$$i_{\text{dis}} = i_{\text{dis1}} + i_{\text{dis2}} = C_0 \frac{\text{d}}{\text{d}t}\left[\frac{1}{2}v_{\text{a}} + \frac{1}{2}v_{\text{b}} - \left(\frac{1}{2}v_{\text{c}} + \frac{1}{2}v_{\text{d}}\right)\right] \tag{9.18}$$

The general formula for the total displacement current flowing through transformer inter-winding capacitance has been derived in Chap. 6, expressed as

$$i_{\text{dis}} = C_0 \frac{\text{d}}{\text{d}t}\{\lambda_{\text{P}} v_{\text{a}} + (1 - \lambda_{\text{P}}) v_{\text{b}} - [\lambda_{\text{S}} v_{\text{c}} + (1 - \lambda_{\text{S}}) v_{\text{d}}]\} \tag{9.19}$$

By comparing (9.18) and (9.19), the winding structural parameter λ_{P} and λ_{S} are both equal to 0.5. Referring to the two-capacitor expressions in (8.8) for the transformer with a center-tap, C_{ae} and C_{be} are given by

$$\begin{cases} C_{\text{ae}} = [\lambda_{\text{P}} + n(1 - 2\lambda_{\text{S}})]C_0 \\ C_{\text{be}} = \{1 - [\lambda_{\text{P}} + n(1 - 2\lambda_{\text{S}})]\}C_0 \end{cases} \tag{9.20}$$

By substituting $\lambda_{\text{P}} = \lambda_{\text{S}} = 0.5$ into (9.20), yields

$$C_{\text{ae}} = C_{\text{be}} = 0.5C_0 \tag{9.21}$$

As seen, when $C_{01} = C_{02} = 0.5C_0$, C_{ae} is equal to C_{be}. Furthermore, by substituting (9.21) and $C_{\text{p1}} = C_{\text{p2}}$ into (9.11), $k_{\text{HB\_LLC}}$ is calculated as 0.5, so the turns number of the CMVC windings is half of the primary winding turns.

9.4.4 *Capacitive Coupling Effect*

As the CMVC windings are closely placed with the inductor/transformer winding, the capacitive effect must be considered. Because the capacitive coupling is confined within the circuit and the PE is not involved, this coupling does not create additional CM noise path. However, the circuit operation can be affected if the parasitic capacitance between the CMVC windings and their coupled windings is significantly large. To address this issue, the parasitic capacitance should be kept small, which can be achieved by enlarging the distance between these windings. Consequently, the leakage inductance between the CMVC windings and their coupled windings will be increased. The method for dealing with this leakage inductance will be discussed below.

9.4.5 *Effect of Leakage Inductance*

In this sub-section, the effect of the leakage inductance of the CMVC winding is discussed. The buck converter with the CMVC method is taken as an example, as shown in Fig. 9.12a, where the leakage inductance and magnetizing inductance are included. It is assumed that two CMVC windings are perfectly coupled, and they have equal leakage inductance $L_{\mathrm{lk\_p}}$ with respect to the filtering winding. Following the previous procedure of deriving the CM noise circuit, the CM noise circuit with substituted sources is given in Fig. 9.12b. By simplifying the circuit at the right side of 25-Ω resistor, the Thevenin equivalent circuit is derived in Fig. 9.14c, where

$$v_{\mathrm{ENS}} = \left(\frac{L_{\mathrm{f}}}{L_{\mathrm{f}} + L_{\mathrm{lk\_p}}} \cdot k_{\mathrm{buck}} - \frac{C_{\mathrm{p1}}}{C_{\mathrm{p1}} + C_{\mathrm{p2}} + C_{\mathrm{p3}} + C_{\mathrm{B}} + C_{\mathrm{p4}}} \right) v_{\mathrm{DFW}} \tag{9.22}$$

Referring to Fig. 9.12c and (9.22), the leakage inductance has affect on the equivalent noise source v_{ENS} and the CM noise source impedance. In order to reduce the CM noise, the coefficient before v_{DFW} in (9.22) should be zero, yields

$$k_{\mathrm{buck}} = \left(1 + \frac{L_{\mathrm{lk\_p}}}{L_{\mathrm{f}}}\right) \cdot \frac{C_{\mathrm{p1}}}{C_{\mathrm{p1}} + C_{\mathrm{p2}} + C_{\mathrm{p3}} + C_{\mathrm{B}} + C_{\mathrm{p4}}} \tag{9.23}$$

According to Fig. 9.12a, the leakage inductance and the magnetizing inductance form a voltage divider, and the induced voltages across the CMVC windings are actually smaller. To counteract the effect of the leakage inductance, k_{buck} is increased accordingly by comparing (9.23) and (9.15).

9.5 Experimental Verification

In this Section, the prototypes of one non-isolated DC-DC converter (buck converter) and one isolated DC-DC converter (half-bridge *LLC* resonant converter) are built and tested in the lab, for verifying the effectiveness of the CMVC method. The CM parasitic capacitance is measured by a LCR meter (Chroma 3302) at 200-kHz test frequency.

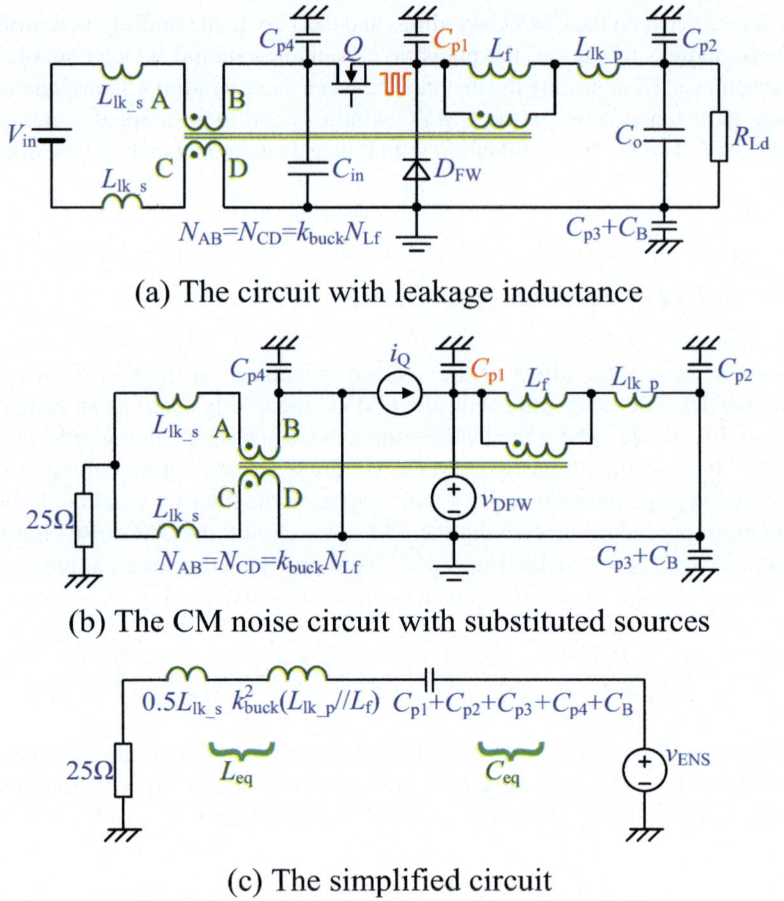

Fig. 9.12 Buck converter with the CMVC method considering leakage inductance

9.5.1 Buck Converter

The buck converter with synchronous rectification (see Fig. 9.4) is operated at 200 kHz. Table 9.1 lists the main parameters. Because the upper and bottom switches are both mounted on a heatsink which is connected to the PE, the dominant parasitic capacitance between the circuit nodes and PE are C_{p1} and C_{p4} (37 pF), and C_{p2} and C_{p3} are neglected. The output filter inductor adopts PQ 26/25 core, and all the windings use single strand enameled wire. The diameter for the inductor and the CMVC winding are 1.5 mm and 0.7 mm, respectively.

For realizing the CMVC method, k_{buck} is chosen as 0.5. Therefore, the CMVC winding takes 7 turns, which is half of the inductor winding turns. The measured parasitic capacitance between the CMVC windings and inductor winding is 23 pF, which is negligible for the buck converter operating at 200 kHz. Figure 9.13 shows

Table 9.1 Related parameters of the buck converter

Input voltage V_{in}	48 V	Filter inductance L_f	28 μH
Output voltage V_o	12 V	Output capacitance C_f	20 μF
Output power P_o	100 W	Capacitance C_{p1}	37 pF
Input capacitance C_{in}	10 μF	Capacitance C_{p4}	37 pF

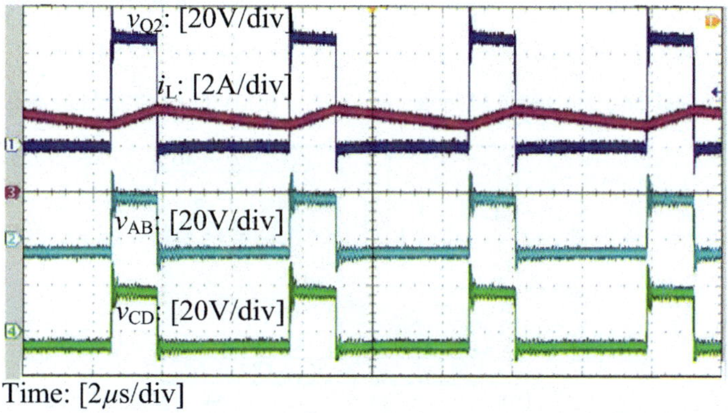

Fig. 9.13 Experimental waveforms of buck converter adopting the CMVC method

the measured waveforms at full-load. As seen, the buck converter works normally and the voltage across the CMVC winding is half of the drain-to-source voltage v_{Q2} of the bottom switch. In addition, the reflected CM current in the inductor current is too small to be noticed.

Figure 9.14 shows the comparison of the CM EMI spectra of the buck converter without CM noise reduction, with CM inductor and with the CMVC method. In Fig. 9.14a, the original CM noise of the buck converter is compared with the buck converter using a CM inductor of 6.3 μH, which is close to the magnetizing inductance of the CMVC windings. At low frequencies, since the impedance of this CM inductor is very small compared with the impedance of the CM parasitic capacitance, the CM noise reduction is not obvious. However, at high frequencies, the impedance of this CM inductor exceeds the impedance of parasitic capacitance, and significant reduction can be observed.

Figure 9.14b shows the comparison of the CM noise of the buck converter with the CMVC method and with the CM inductor for fair comparison. Referring to Fig. 9.12a, by adding the CMVC windings, $L_f = 28$ μH, $L_{lk\_p} = 1.44$ μH, and $L_{lk\_s} = 0.60$ μH. Because L_{lk_p} is far smaller than L_f, the compensation voltage is close to the ENS, and 12-dB reduction can be observed in Fig. 9.14b when the CMVC method is applied. By substituting the related parameters into (9.23), the balancing

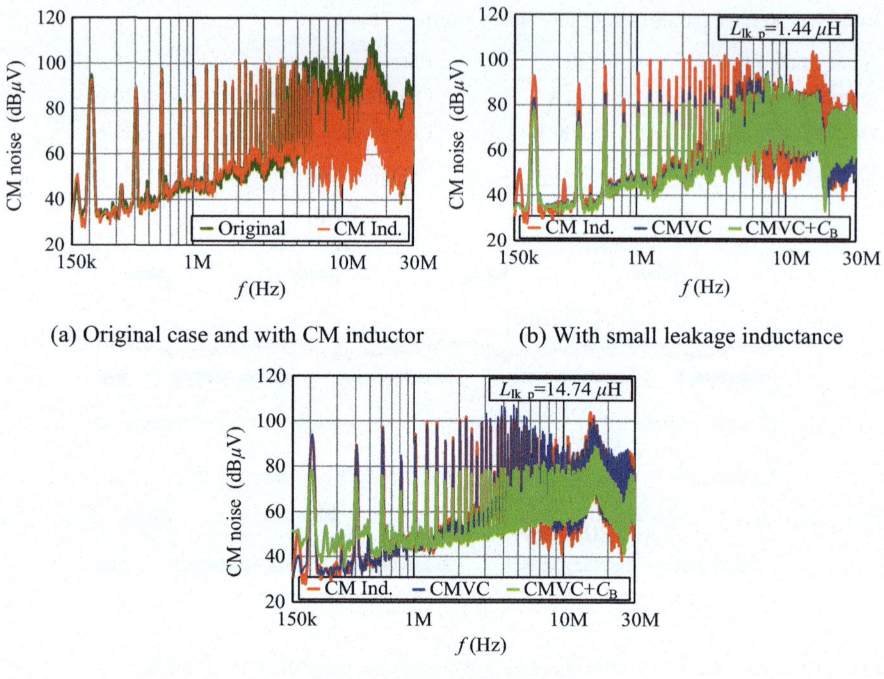

(a) Original case and with CM inductor

(b) With small leakage inductance

(c) With large leakage inductance

Fig. 9.14 Comparison of the CM noise under various conditions

capacitor C_B is calculated as 3.8 pF. After adding this balancing capacitor to be connected between the power ground and PE, the CM noise is further reduced by 5 dB, which validates the analysis in Sect. 9.4.5.

In order to further validate the analysis of the case with large leakage inductance, an external inductor (13.30 μH) is added to be in series with the inductor winding, and a CM inductor (2.90 μH) is added in series with the CMVC winding. Although the effective filter inductance is increased, the duty cycle stays the same since the buck converter operates in continuous conduction mode (CCM), and the spectra of v_{Q2} are not changed. As seen in Fig. 9.14c, there is no obvious reduction at low frequencies when the CMVC method is applied, because the compensation voltage becomes lower in this case, which does not match the ENS of the buck converter. By substituting the related parameters into (9.23), C_B is calculated as 39.0 pF. By adding this balancing capacitor connected between the power ground and the PE, the resultant CM noise is significantly reduced at low frequencies. The noise reduction at high frequencies is still not known, which requires more detailed modeling with other parasitic components in the circuit.

9.5.2 *Half-Bridge* LLC *Resonant Converter*

Figure 9.15 shows the HB *LLC* resonant converter prototype and the transformer. The auxiliary winding is one turn and is connected in the primary circuit for powering the driving and control circuit. Because the primary winding acts as shielding between the auxiliary winding and secondary winding, the auxiliary winding does not affect the CM noise.

Table 9.2 lists the related parameters. The circuit with the CMVC method is given in Fig. 9.10d. The CMVC winding uses single strand enameled wire, because most of the AC components from the input current are filtered by the input capacitor. The resonant inductor uses 0.1 mm × 150 litz wire, and has 10 turns in a PQ 26/20 core.

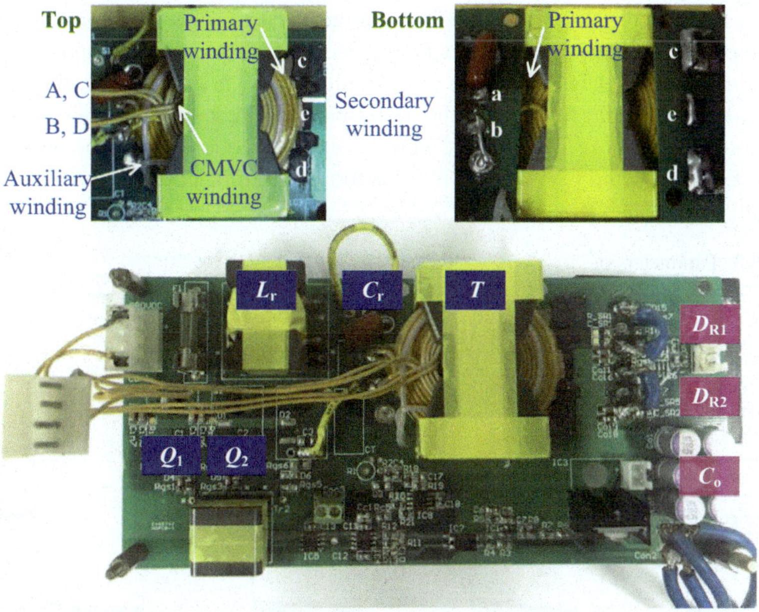

Fig. 9.15 Photo of the HB *LLC* resonant converter protptype

Table 9.2 Main circuit parameters of the HB *LLC* resonant converter

Input voltage V_{in}	385 V	Resonant inductance L_r	13.5 μH
Output voltage V_o	48 V	Resonant capacitance C_r	57 nF
Output power P_o	1 kW	Magnetizing inductance L_m	86 μH
Resonant frequency f_r	180 kHz	Input capacitance C_{in}	2.5 μF
Turns ratio n	1/4	Output capacitance C_o	600 μF

The transformer winding structure is shown in Fig. 9.11, and the winding dimensions are listed in Table 9.3. Since the same type of insulation material with equal thickness is used between Pri1 and Sec1-1, as well as Pri2 and Sec2-2, C_{01} and C_{02} are nearly the same, and they are measured as 53 pF. Likewise, C_{p1} and C_{p2} are also equal, and their measured values are both 20 pF. According to the analysis in Sect. 9.4.3, the turns N_{AB} (N_{CD}) of the CMVC windings are half of the primary winding turns, which is 4. Besides, the measured parasitic capacitance between the CMVC windings and the primary winding is 26 pF, which is negligible for the *LLC* resonant converter operating around 170.35 kHz.

Figure 9.16 shows the experimental waveforms of the *LLC* resonant converter at full-load, including the primary winding voltage v_{ab}, the resonant inductor current i_{Lr}, and the voltage v_{res} across the resonant branch consisted of L_r and C_r. As seen, the switching frequency is around 170.35 kHz. In addition, v_{res} presents impulse scheme, because the slope of the resonant inductor current has sudden change around the intersection of two-component resonance and three-component resonance, as highlighted in Fig. 9.16. Since the pulse width of v_{res} is quite small, the harmonic amplitude of v_{res} is far lower than the harmonic amplitude of v_{ab} at low frequencies. Therefore, the effect of v_{res} on the CM noise at low frequencies can be neglected. In addition, the reflected CM current in i_{Lr} is too small to be noticed.

Table 9.3 Parameters of power transformer

	Winding	Dimension	Turns
Primary winding ab	Litz wire	0.1 mm × 80	Pri1: 8
			Pri2: 8
Secondary winding cd	PCB winding	4 oz, 9.75 mm	4

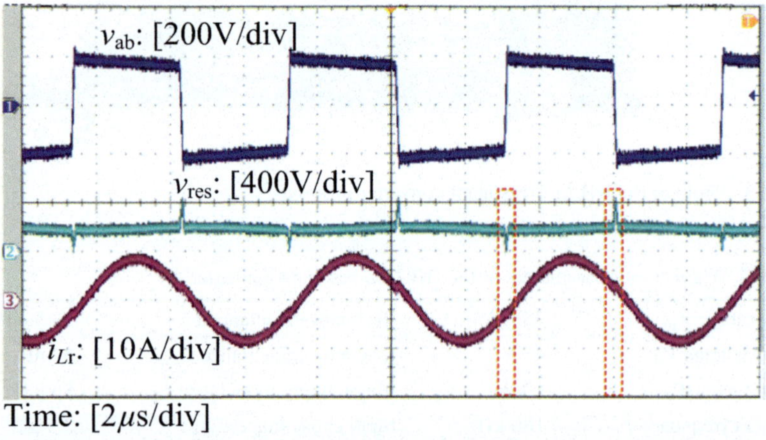

Fig. 9.16 Measured voltage across the primary winding and the resonant branch

Figure 9.17 shows the experimental waveforms of the *LLC* resonant converter with the CMVC method, including the primary winding voltage v_{ab} and the voltage v_{AB}, v_{CD} across the CMVC windings. As seen, these voltage waveforms are similar, and v_{ab} is twice the amplitude of v_{AB} and v_{CD}.

Figure 9.18 compares the measured CM noise of the HB *LLC* resonant converter with and without the CMVC method, as well as the CM EMI of the converter with CM inductor. The CM inductance is 22.30 μH, which is close to the magnetizing inductance of the CMVC windings. As seen, when the CM inductor is inserted, the CM noise does not change at low frequencies and is reduced at high frequencies, which is similar with the results of buck converter. By applying the CMVC method, the CM EMI is reduced by nearly 20 dB, which validates the effectiveness of the proposed method.

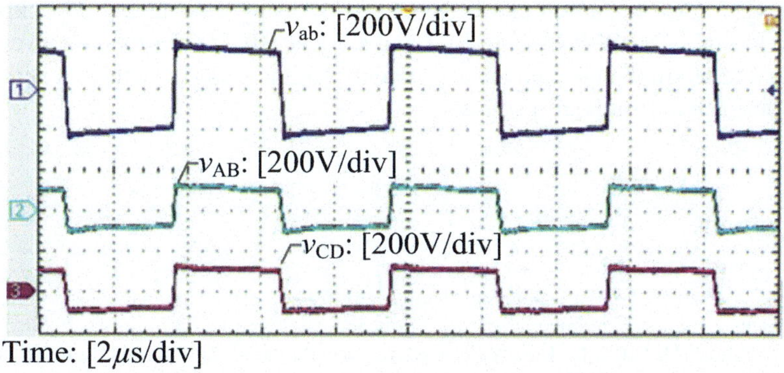

Fig. 9.17 Measured voltage across the primary winding of the transformer and the winding for the CM noise reduction

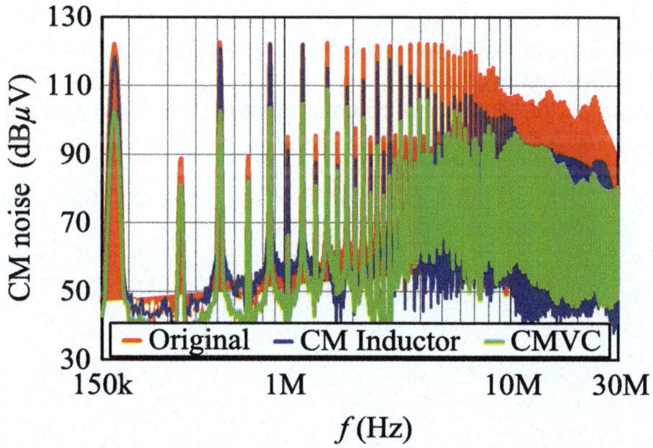

Fig. 9.18 Comparison of the original CM noise, the CM noise with CM inductor and the CM noise with CMVC method at full-load

9.6 Summary

In this chapter, the CMVC method is proposed to reduce the CM noise of DC-DC converters. By adding two windings connected between the input voltage source and the converter input side, and coupling them with the inductor or transformer winding, a compensation voltage is generated for cancelling the CM noise of DC-DC converters. The CMVC method can be implemented in both non-isolated and isolated DC-DC converters. Specifically, for boost, Cuk and SEPIC converters, the CMVC method can be achieved by dividing the input inductor winding into two coupled windings placed at the input positive line and the input return line; and for flyback and forward converters, the primary winding can be split into two coupled windings in series with the main switch, respectively. The CMVC method is recommended to be applied in the converter with low input current, and the effect of leakage inductance on the CM noise reduction can be managed by using a balancing capacitor. The prototypes of a 100-W buck converter and a 1-kW HB *LLC* resonant converter are built and tested in the lab, and the effectiveness of the CMVC method is verified by the experimental results, respectively.

References

1. L. Xie, X. Ruan, Z. Ye, Equivalent noise source: an effective method for analyzing common mode noise in isolated power converters. IEEE Trans. Industr. Electron. **63**(5), 2913–2924 (2016)
2. D. Cochrane, D. Chen, D. Boroyevic, Passive cancellation of common-mode noise in power electronic circuits. IEEE Trans. Power Electron. **18**(3), 756–763 (2003)
3. S. Ogasawara, H. Ayano, H. Akagi, An active circuit for cancellation of common mode voltage generated by a PWM inverter. IEEE Trans. Power Electron. **13**(5), 835–841 (1998)
4. S. Wang, P. Kong, F.C. Lee, Common mode noise reduction for boost converters using general balance technique. IEEE Trans. Power Electron. **22**(4), 1410–1416 (2007)
5. H. Zhang, S. Wang, Y. Li, Q. Wang, D. Fu, Two-capacitor transformer winding capacitance models for common- mode EMI noise analysis in isolated dc−dc converters. IEEE Trans. Power Electron. **32**(11), 8458–8469 (2017)
6. L. Xie, X. Ruan, and Z. Ye, Modeling of common mode noise in phase-shift full-bridge converter, in *Proceedings of IEEE Annual Conference of Industrial Electronics Society (IECON)* (2016), pp. 1371–1375
7. L. Xie, X. Ruan, Z. Ye, Reducing common mode noise in phase-shift full-bridge converter. IEEE Trans. Industr. Electron. **65**(10), 7866–7877 (2018)
8. P. Kong, S. Wang, F.C. Lee, Z. Wang, Reducing common-mode noise in two-switch forward converter. IEEE Trans. Power Electron. **26**(5), 1522–1533 (2011)

Chapter 10
Reducing the Common-Mode Currents at the Input and Output Sides in Non-isolated DC-DC Conveters

10.1 The Common-Mode (CM) Noise Model of Buck Converter Considering the CM Impedances at the Input and Output Sides

In non-isolated DC-DC converters, there are high dv/dt nodes and parasitic capacitances between these nodes and the protective earth (PE). The displacement currents flowing through these parasitic capacitances are generated, leading to the common-mode (CM) noise. The CM noise not only appears at the input side, but also propagates into the output side, affecting the operation of loads. In this section, the CM noise model of buck converter will be derived with the consideration of the CM impedances at both the input and output sides.

Figure 10.1 shows the buck converter and its CM noise propagation path, where, Q is the power switch, D_{FW} is the freewheeling diode, L_{f} is the filter inductor, C_{o} is the output filter capacitor, and R_{Ld} is the load. C_{pA}, C_{pB}, C_{pC}, C_{p1} and C_{p2} are the CM parasitic capacitaces. The line impedance stabilization network (LISN) is placed between the input voltage source and the input side of the converter for providing specific impedance for the CM noise current [1].

The derivation of the CM noise circuit of the buck converter is illustrated as follows. Owing to the high dv/dt characteristics of node B, D_{FW} is replaced with a voltage source that replicates its waveform. The capacitors C_{in} and C_{o} are treated as short circuit because their dv/dt are much smaller than that of the voltage across D_{FW}. For avoiding voltage loops, switch Q and inductor L_{f} are replaced with current sources that replicate their waveforms [2]. Finally, the CM impedance at the LISN side can be simplified as a 25-Ω resistor, and the CM impedance at the output side is essentially the parallel connection of C_{p1} and C_{p2}. The resultant CM noise equivalent circuit with substituted sources is given in Fig. 10.2.

© The Author(s), under exclusive license to Springer Nature Singapore Pte Ltd. 2024

X. Ruan et al., *Conducted Electromagnetic Interference in Power Converters: Modeling, Prediction and Reduction*, CPSS Power Electronics Series,
https://doi.org/10.1007/978-981-99-9295-9_10

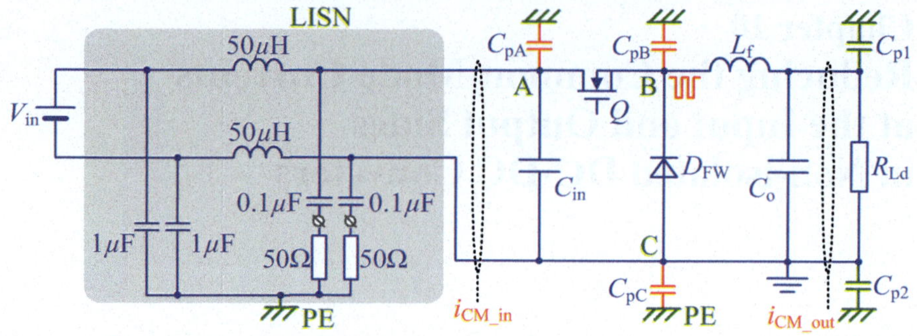

Fig. 10.1 Buck converter and the CM noise propagation paths

Fig. 10.2 The CM noise equivalent circuit with substituted sources

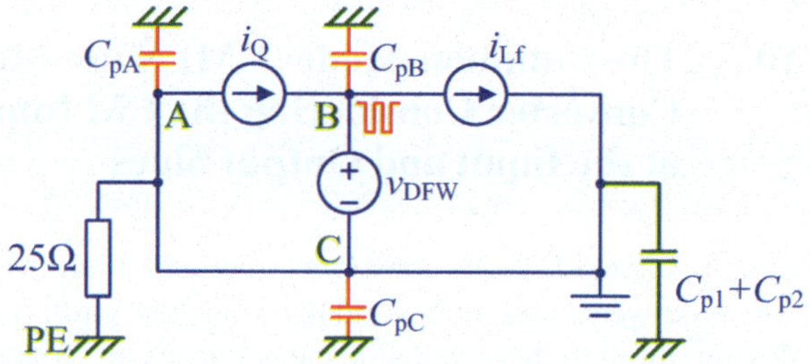

Superposition theorem is adopted to simplify the equivalent circuit in Fig. 10.2. Firstly, the voltage source v_{DFW} is shorted to only consider the excitation of current sources i_Q and i_{Lf}. Since i_Q and i_{Lf} are essentially shorted, they do not cause CM noise. Secondly, by opening the current sources i_Q and i_{Lf}, the resultant CM noise model is given in Fig. 10.3a.

By using Thevenin's theorem, the CM noise model in Fig. 10.3a can be simplified as Fig. 10.3b, where, v_{ENS} is the equivalent noise source (ENS) [3], and C_{sum} is the total parasitic capacitance. v_{ENS}, C_{sum} and C_{RL} are expressed as

$$v_{ENS} = k_{buck} v_{DFW} \tag{10.1}$$

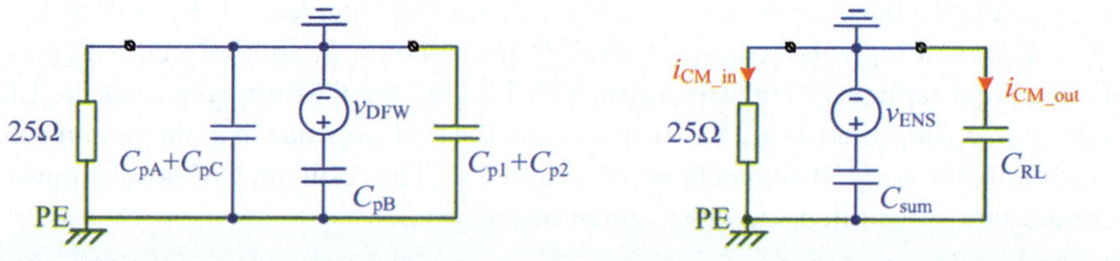

(a) The CM noise model with v_{DFW} (b) The CM noise model with v_{ENS}

Fig. 10.3 Simplification of the CM noise equivalent circuit

$$C_{\mathrm{sum}} = C_{\mathrm{pA}} + C_{\mathrm{pB}} + C_{\mathrm{pC}} \tag{10.2}$$

$$C_{\mathrm{RL}} = C_{\mathrm{p1}} + C_{\mathrm{p2}} \tag{10.3}$$

where, the coefficient k_{buck} is given by

$$k_{\mathrm{buck}} = C_{\mathrm{pB}}/C_{\mathrm{sum}} \tag{10.4}$$

As the voltages of D_{FW} and L_{f} have the same AC component [4], the ENS of the buck converter can also be written as

$$v_{\mathrm{ENS}} = k_{\mathrm{buck}} v_{Lf} \tag{10.5}$$

10.2 The CM Current Reduction of Both Sides in Buck Converters

10.2.1 Derivation of the CM Noise Cancellation Method

According to Fig. 10.3b, to eliminate the CM currents $i_{\mathrm{CM\_in}}$ and $i_{\mathrm{CM\_out}}$, two voltage sources v_{CM1} and v_{CM2} are added, as shown in Fig. 10.4a. When $v_{\mathrm{CM1}} = v_{\mathrm{CM2}} = v_{\mathrm{ENS}}$, the positive terminals of these voltage sources are equi-potential, so they can be shorted, and the circuit shown in Fig. 10.4a can be equivalently transformed into Fig. 10.4b. As seen, no excitation exists, and thus the CM currents $i_{\mathrm{CM\_in}}$ and $i_{\mathrm{CM\_out}}$ are both eliminated.

Referring to Chap. 9, v_{CM1} and v_{CM2} can be are implemented by the CM windings coupled with the inductor L_{f}, as shown in Fig. 10.5. Since the added CM windings do not affect input and output voltages, the operation of the buck converter with coupled windings is identical to the original buck converter.

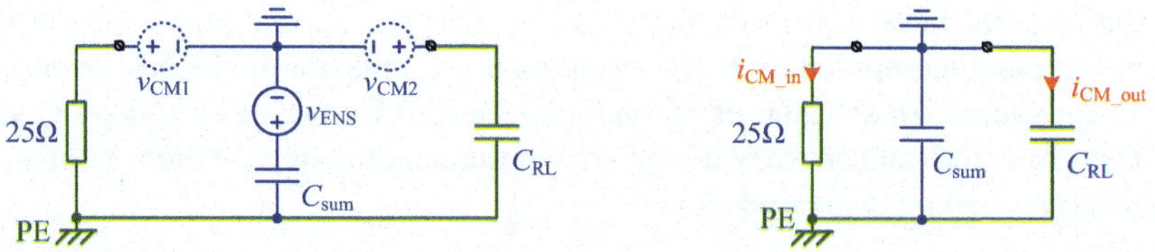

(a) Adding two compensation voltage source (b) The circuit with cancelled source

Fig. 10.4 The principle of CM voltage reduction at both sides

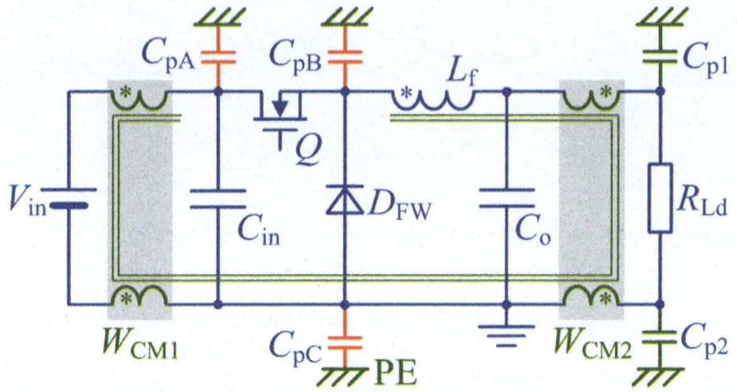

Fig. 10.5 The realization of CM noise reduction at both sides

Referring to Fig. 10.4a, the turns N_{CM} of each winding in W_{CM1} and W_{CM2} are the same, which is equal to

$$N_{CM} = k_{buck} N_{Lf} \tag{10.6}$$

where, N_{Lf} is the turns of inductor winding L_f.

When the leakage inductance of the windings in W_{CM2} is negilible, the voltage of C_o equals to the voltage of R_{Ld}. So, C_o can be moved to be connected in parallel with R_{Ld}, as shown in Fig. 10.6a. Then, the inductor winding of L_f and the upside winding in W_{CM2} can be combined into a single one with $(1–k_{buck})N_{Lf}$ turns, as shown in Fig. 10.6b.

Furthermore, since the downside windings in Fig. 10.6b have the same turns, the dotted terminals in these two windings have the same electric potential, and thus these two dotted terminals can be connected together. By paralleling these two windings, one set of winding can be saved, and the simplified circuit is given in Fig. 10.6c.

10.2.2 Comparison of the Total Cross-Sectional Area of Windings

For the original buck converter, by assuming that the current ripple of the filter inductor is small compared to its DC component I_{out} (the nominal load current), the root mean square (rms) value of the inductor current is very close to its DC value [5]. Therefore, the total cross-sectional area of the winding in L_f, which occupies the core window, can be expressed as

$$A_{W\_Lf} = N_{Lf} I_{out} / J \tag{10.7}$$

where, J is the current density. As seen, A_{W_Lf} is proportional to the ampere-turns of L_f.

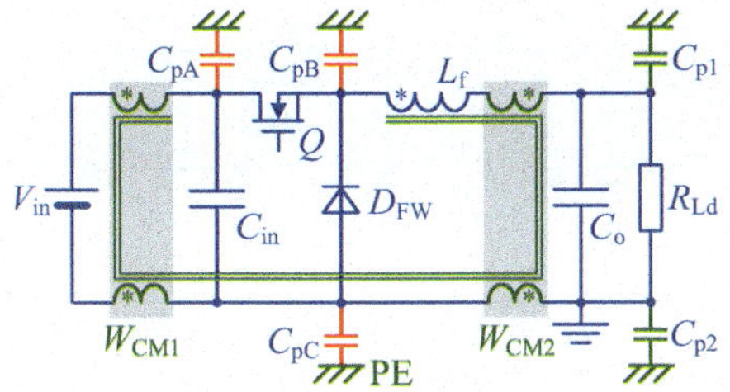

(a) Moving C_o to the right side

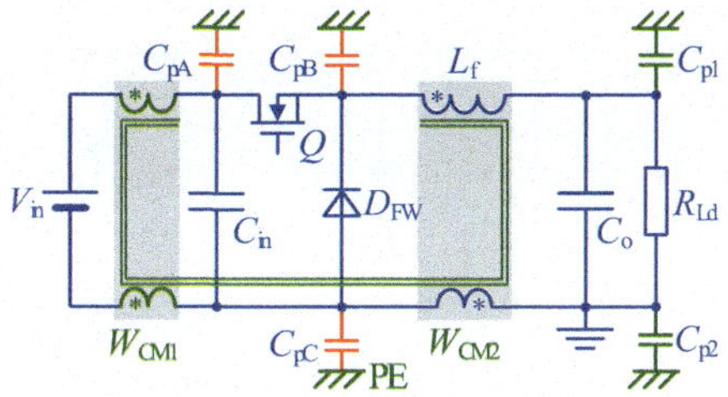

(b) Merging two winding at high side

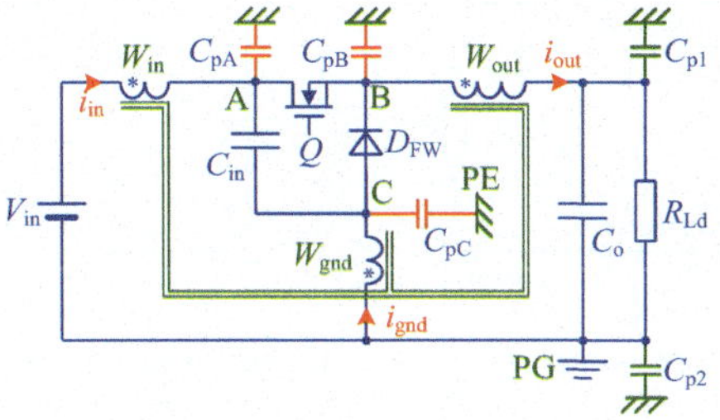

(c) The buck converter with coupled windings

Fig. 10.6 The simplification of Fig. 10.5

For the buck converter with coupled windings in Fig. 10.6c, the total cross-sectional area (A_{Win}, A_{Wgnd} and A_{Wout}) of each winding (W_{in}, W_{gnd} and W_{out}) is also proportional to each ampere-turns, yields

$$A_{Win} = k_{buck} N_{Lf} I_{in} / J \tag{10.8}$$

$$A_{Wgnd} = k_{buck} N_{Lf} (I_{out} - I_{in}) / J \tag{10.9}$$

$$A_{Wout} = (1 - k_{buck}) N_{Lf} I_{out} / J \tag{10.10}$$

where, I_{in} is the input DC current of the converter.

According to (10.8)–(10.10), since the total ampere-turns of the coupled windings are equal to that of L_f, the total cross-sectional area of windings keeps the same.

For a better understanding, Fig. 10.7 shows an intuitive demonstration for the relationship between the buck converter and the buck converter with coupled windings.

Firstly, the inductor winding in Fig. 10.7a is split into two coupled windings by dividing the turns, as shown in Fig. 10.7b, where, the ampere-turns of W_{out} and W_{ig} are $(1–k_{buck})N_{Lf}I_{out}$ and $k_{buck}N_{Lf}I_{out}$, respectively. Secondly, W_{ig} is split into two paralleled windings W_{in} and W_{gnd} by dividing the cross-sectional area, as shown in Fig. 10.7c, where, the ampere-turns of W_{in} and W_{gnd} are $k_{buck}N_{Lf}I_{in}$ and $k_{buck}N_{Lf}(I_{out} - I_{in})$, respectively.

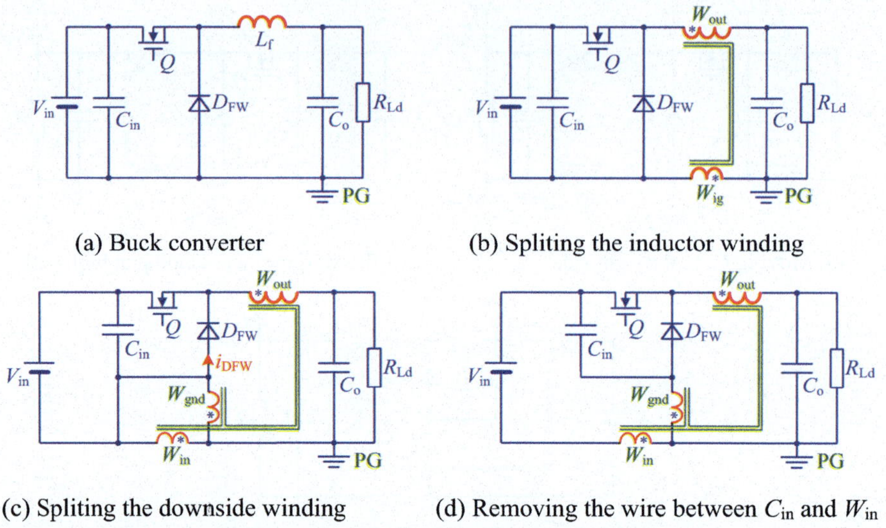

(a) Buck converter

(b) Spliting the inductor winding

(c) Spliting the downside winding

(d) Removing the wire between C_{in} and W_{in}

Fig. 10.7 The demonstration of the buck converter with coupled windings

Referring to Fig. 10.7c, since W_{in} and W_{gnd} have the same voltage, either the wire between C_{in} and W_{in} or the wire between C_{in} and W_{gnd} can be open. Since i_{DFW} is a pulsating current, the coupled winding should not be connected in series with D_{FW} due to the leakage inductance of W_{gnd}. Therefore, the wire between C_{in} and W_{in} can be open, as shown in Fig. 10.7d. Finally, to make the input and output share the same ground, W_{in} can be moved to the positive terminal of V_{in}, which is the buck converter with coupled windings in Fig. 10.6c. As seen in Fig. 10.7, the ampere-turns of the inductor winding of buck converter is essentially split, so this method is named as the split-winding (SW) method, and the converter is the SW buck converter.

10.3 The Balance Condition Considering Imperfect Coupling

For the SW buck converter in Fig. 10.6c, three windings are coupled and their coupling would not be perfect in practical applications. As a result, the CM noise reduction can be influenced. In this section, the effect of imperfect coupling is discussed.

To derive the CM noise model for Fig. 10.6c, the circuit with substituted sources is given in Fig. 10.8a. By simplifying the circuit in Fig. 10.8a, the resultant circuit is given in Fig. 10.8b, where, W_{ig} is the parallel of W_{in} and W_{gnd}. As long as the bridge in the middle is balanced, the CM currents at the input and output sides are zero accordingly.

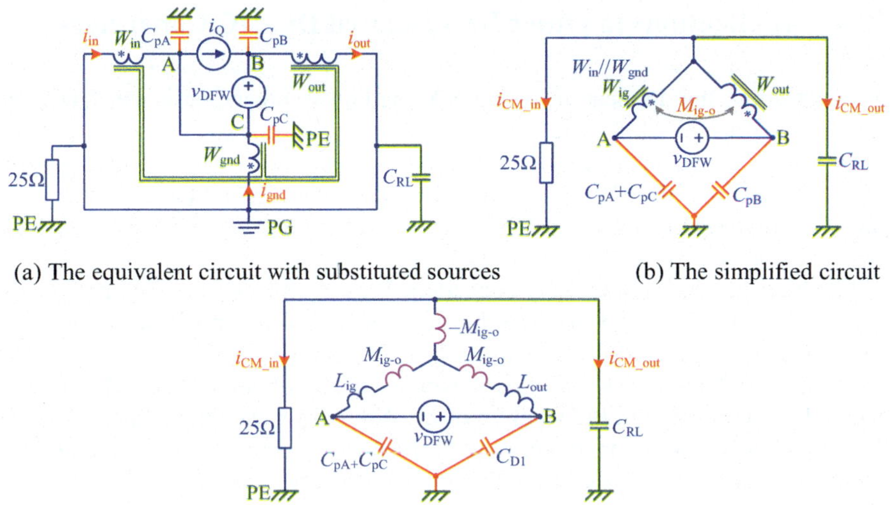

(a) The equivalent circuit with substituted sources

(b) The simplified circuit

(c) The decoupled CM noise model

Fig. 10.8 Split-winding buck converter considering practical coupling

To derive the balance condition for the circuit in Fig. 10.8b, the windings W_{ig} and W_{out} are decoupled, and the circuit is given in Fig. 10.8c. The balance condition is

$$\frac{C_{pB}}{C_{pA}+C_{pC}} = \frac{L_{ig}+M_{ig\text{-}o}}{L_{out}+M_{ig\text{-}o}} \tag{10.11}$$

where, L_{ig} and L_{out} are the self inductance of W_{ig} and W_{out}, respectively, and $M_{ig\text{-}o}$ is the mutual inductance between W_{ig} and W_{out}.

In practical applications, C_{pA} and C_{pB} are formed between the switches and the heatsink, and they are typically larger than C_{pC}, so C_{pA} and C_{pB} are dominant and C_{pC} is treated as 0. In addition, the same type of insulation material with equal thickness is normally used between the switches and the heatsink, so C_{pA} and C_{pB} are nearly equal. According to (10.4), when $C_{pA} = C_{pB}$, the coefficient k_{buck} is equal to 0.5, and the turns of W_{in}, W_{gnd} and W_{out} are equal.

For a compact design of magnetic component in the SW buck converter, W_{in}, W_{gnd} and W_{out} should be wound in the same core to achieve tight coupling. As a result, the self inductance L_{ig} and L_{out} and the mutual inductance $M_{ig\text{-}o}$ are nearly the same. By substituting $C_{pA} = C_{pB}$, $C_{pC} = 0$ and $L_{ig} = L_{out} = M_{ig\text{-}o}$ into (10.11), the balance condition for reducing the CM noise is satisfied, so extra capacitors are not required to achieve balance.

10.4 Applications in Other Non-isolated DC-DC Converters

In this section, the application of the SW method in other non-isolated DC-DC converters is given below. The SW boost converter can be directly derived from the SW buck converter, since their power flow direction is reversed to each other. Figure 10.9a shows the SW boost converter, where, the winding turns of W_{in}, W_{gnd} and W_{out} are listed in Table 10.1, N_{Lb} is the turns of original boost inductor winding, and the coefficient k_{bst} is the same as (10.4).

For other non-isolated DC-DC converters, the circuit using the SW method are listed in Fig. 10.9b–e, and the winding turns of W_{in}, W_{gnd} and W_{out} are listed in Table 10.1.

For the SW buck-boost converter shown in Fig. 10.9b, the added CM windings cannot be combined with the winding of L_{b}. As a result, the total ampere-turns of the coupled windings are increased, and the total cross-sectional area of the coupled windings is larger than that of L_{b} in the original buck-boost converter.

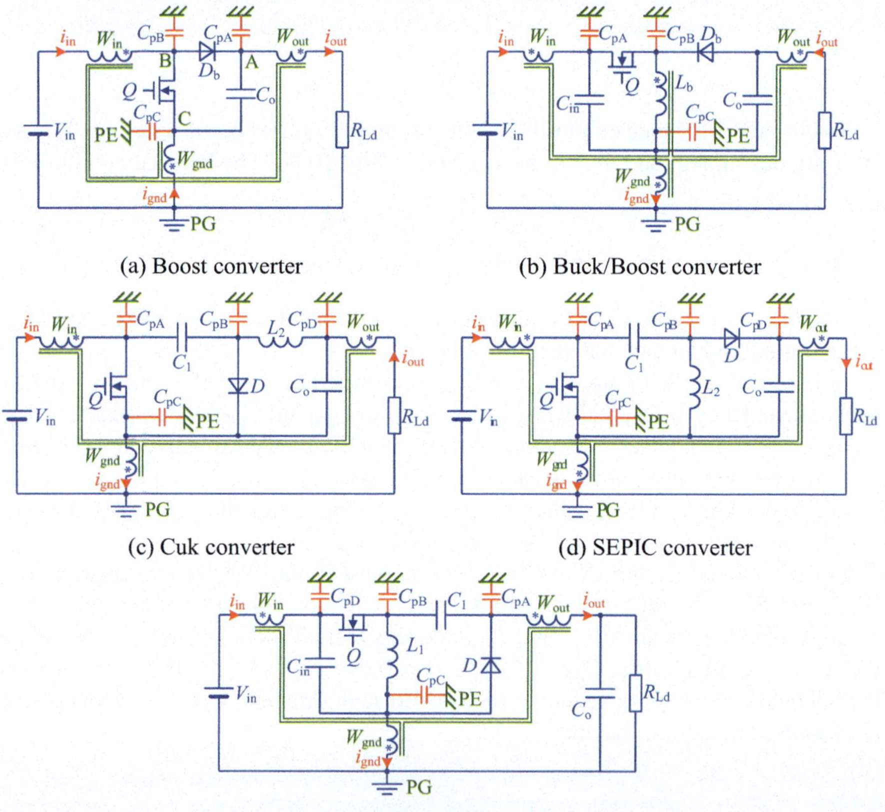

(a) Boost converter
(b) Buck/Boost converter
(c) Cuk converter
(d) SEPIC converter
(e) Zeta converter

Fig. 10.9 Application of the SW configuration in basic non-isolated DC-DC converters

Table 10.1 Winding turns in basic non-isolated DC-DC converters and their SW ones

Topology	Inductor and its winding turns	The SW converter			
		W_{in}	W_{gnd}	W_{out}	Coefficient
Buck	L_f: N_{Lf}	$k_{buck}N_{Lf}$	$k_{buck}N_{Lf}$	$(1-k_{buck})N_{Lf}$	k_{buck}, k_{bst}, $k_{bb} = \frac{C_{pB}}{C_{pA}+C_{pB}+C_{pC}}$
Boost	L_b: N_{Lb}	$(1-k_{bst})N_{Lb}$	$k_{bst}N_{Lb}$	$k_{bst}N_{Lb}$	
Buck-boost	L_b: N_{Lb}	$k_{bb}N_{Lb}$	$k_{bb}N_{Lb}$	$k_{bb}N_{Lb}$	
Cuk	L_1: N_{L1}	$(1-k_{Cuk})N_{L1}$	$k_{Cuk}N_{L1}$	$k_{Cuk}N_{L1}$	k_{Cuk}, k_{SEPIC}, $k_{Zeta} = \frac{C_{pA}+C_{pB}}{C_{pA}+C_{pB}+C_{pC}+C_{pD}}$
SEPIC	L_1: N_{L1}	$(1-k_{SEPIC})N_{L1}$	$k_{SEPIC}N_{L1}$	$k_{SEPIC}N_{L1}$	
Zeta	L_1: N_{L2}	$k_{Zeta}N_{L2}$	$k_{Zeta}N_{L2}$	$(1-k_{Zeta})N_{L2}$	

For the SW Cuk converter shown in Fig. 10.9c, since the average current in W_{gnd} is larger than the input current, the total ampere-turns of the coupled windings are increased, and the total cross-sectional area of the coupled winding is also larger than that of L_1 in the original Cuk converter.

For the SW SEPIC converter shown in Fig. 10.9d, the winding of inductor L_1 near the input voltage source is split into coupled windings W_{in}, W_{gnd} and W_{out}. When the SW SEPIC converter operates in voltage step-up mode, we have $I_{in} > I_{out}$. Therefore, the total ampere-turns of the coupled windings are equal to that of L_1, and the total cross-sectional area of the coupled windings is the same as that of L_1 in the original SEPIC converter. However, when the SW SEPIC converter operates in voltage step-down mode, $I_{in} < I_{out}$, the total ampere-turns of the coupled windings are increased, and so does the total cross-sectional areas.

For the Zeta converter, since its power flow direction is reverse to that in the SEPIC converter, the SW Zeta converter is directly obtained by reversing the SW SEPIC converter, as shown in Fig. 10.9e. As the duality, for the SW Zeta converter working in voltage step-down mode, the total cross-sectional area of the coupled windings is the same as that of L_2 in the original Zeta converter. For the SW Zeta converter working in voltage step-up mode, the total cross-sectional area of the coupled windings is increased.

10.5 Experimental Verification

To verify the effectiveness of the SW method, a buck converter and a SW buck converter are built, and the CM currents at the input and output sides are measured and compared.

Figure 10.10a shows the SW buck converter prototype, and Fig. 10.10b, c show the inductor in the SW buck converter and the inductor in the buck converter, respectively. For a fair comparison, the buck converter and the SW buck converter use the printed

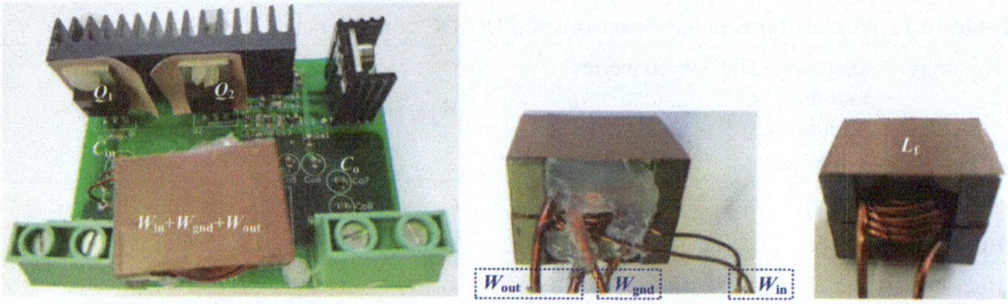

(a) Buck converter prototype (b) SW inductor (c) Original inductor

Fig. 10.10 Photo of the prototype

circuit board (PCB) with the same dimension. In these converters, synchronous rectification is adopted to reduce the conduction loss, and the upper and bottom switches are Q_1 and Q_2, respectively. The switching frequency is 200 kHz.

Table 10.2 lists the main circuit parameters and the parasitic capacitances of the buck converter, where, C_{pA} and C_{pB} are measured by the impedance analyzer (WAYNE KERR 6500B) at the test frequency of 100 kHz, and the switches are disconnected from the circuit during the measurement.

For the magnetic component in the buck and SW buck converter, Table 10.3 lists the winding dimension and measured parameters. Single strand enameled wire is used for the windings in these inductors. The detailed arrangement and connection of the windings are shown in Fig. 10.11. The winding W_{in} is connected between $C_{in}(+)$ and $V_{in}(+)$, which are the positive terminals of C_{in} and V_{in}, respectively. The winding W_{gnd} is placed between $C_{in}(-)$ and $V_{in}(-)$, which are the negative terminals of C_{in} and V_{in}, respectively. The winding W_{out} interconnects the source of Q_1 and the positive terminal of C_o, whose connection is the same as the inductor winding in the buck converter.

Figure 10.12 shows the operation waveforms of the SW buck converter, including the electric potential of the source of switch Q_1, Q_2 with respect to the power ground, and the current flowing through the coupled windings. As seen in Fig. 10.13a, the current flowing through W_{out} is identical to the inductor current in the buck converter. Moreover, the electric potential of the observed nodes are complementary, which enables the CM noise cancellation for the SW method.

Table 10.2 Main circuit parameters and parasitic capacitances of the buck converter

Input voltage V_{in}	48 V	Input capacitance C_{in}	180 μF	Capacitance C_{pA}	27 pF
Output voltage V_o	12 V	Filter inductance L_f	25 μH	Capacitance C_{pB}	27 pF
Load resistance R_L	1.5 Ω	Output capacitance C_o	180 μF	Capacitance C_{pC}	0

Table 10.3 Winding dimension and measured parameters of the inductors

	Original inductor	Split-winding inductor		
		W_{in}	W_{gnd}	W_{out}
Turns	14	7	7	7
Dimension (mm)	1.5	0.7	1.3	1.5
Self inductance	25.16 μH	Paralleled (W_{ig}): 6.54 μH		6.52 μH
Mutual inductance		W_{ig} and W_{out}: 6.05 μH		
Weight	45 g	44 g		
Volume	27.3 × 19.3 × 21.6 mm (PQ 26/20 core)			

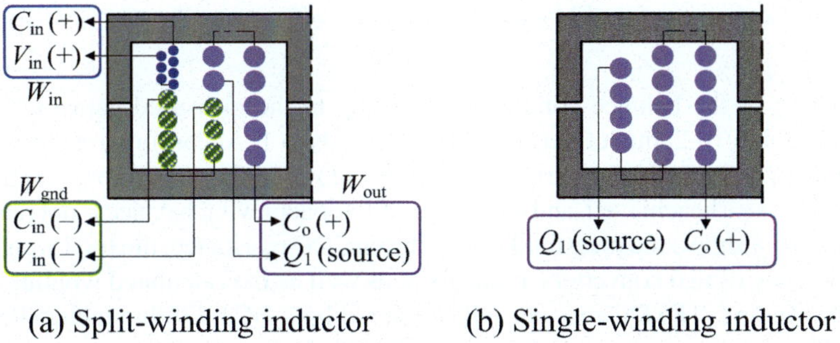

Fig. 10.11 The winding structures of split-winding and single-winding inductor

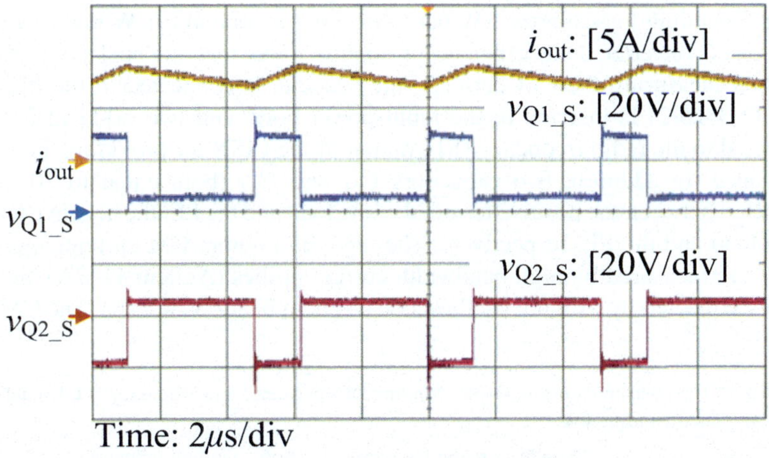

Fig. 10.12 The experimental waveforms of the SW buck converter

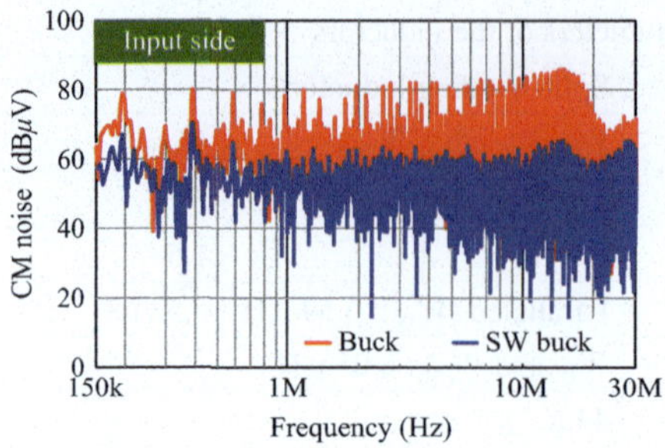

(a) CM noise spectra at the input side

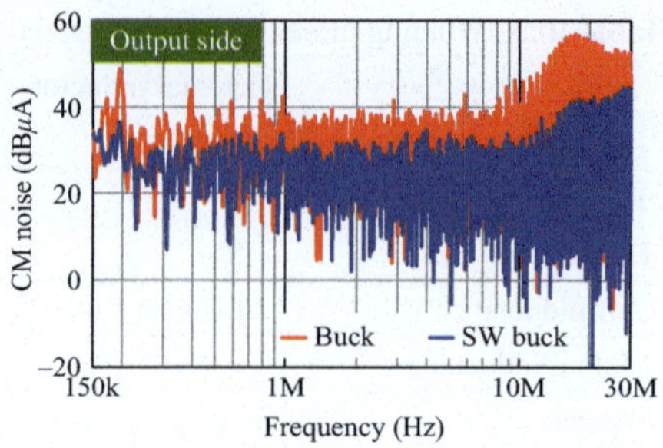

(b) CM noise spectra at the output side

Fig. 10.13 Comparison of the CM noise spectra for the buck and the SW buck converters

In terms of the balance condition, according to the measured parasitic capacitances in Table 10.2, the left term in (10.11) equals to 1. According to the measured inductances in Table 10.3, the right term in (10.11) is 1.0016. Therefore, the balance condition is nearly achieved, and extra capacitor for achieving balance is not required.

Table 10.4 shows the measured ripple current, DC resistance, the total weight and the efficiency of two converters at full-load, as well as the calculated winding losses of each winding. The DC resistance is measured by a DC milliohm meter (RS PRO RM-805), and the winding losses are calculated based on the DC resistance and the DC current of each winding. The other losses for these two converters such as core loss, power semiconductor losses are the same, and it is not compared here. As seen, the total winding losses in the SW buck converter is around 0.1 W higher than that in the buck converter, so its efficiency is slightly lower than the buck converter.

When measuring the CM currents, the heatsink is connected to the PE, LISN (EMCO 3825/2) is inserted in the input power cord, and two 50-Ω resistors are connected to the radio frequency (RF) output of the LISN for providing the specific impedance. In addition, two capacitors C_{p1} and C_{p2} (both equal to 20 pF) are connected between the positive terminal of the load and the PE, the negative terminal of the load and the PE, respectively. The input and output CM currents i_{CM_in} and i_{CM_out} are measured by high bandwidth current probes (Agilent 1147A), and their data are collected by a digital oscilloscope (Agilent MSO-X 3054A). The CM noise

Table 10.4 The measured current ripple, winding DC resistance and efficiency at full-load as well as the calculated winding loss

	Single-winding inductor	Split-winding inductor		
		W_{in}	W_{gnd}	W_{out}
Current ripple (A)	2.00	1.30	0.70	2.00
DC resistance (mΩ)	6.81	24.70	5.88	3.15
Copper loss (mW)	406.92	106.86	197.80	191.65
Efficiency (%)	92.57	92.37		

spectra are then calculated by fast Fourier transformation (FFT) algorithm. During the FFT calculation, a rectangular window with a time length of 100 μs is adopted.

Figure 10.13 compares the CM noise spectrum of these two buck converters. The input side CM current i_{CM_in} is converted into noise voltage spectrum [6, 7] in dBμV by taking into account the voltage across the 50-Ω equivalent resistance at the LISN side.

As seen in Fig. 10.13a, when adopting the SW method, 10- to 20-dB reduction for the input side CM noise is achieved, so it is clear to help reduce the EMI filter size. In Fig. 10.13b, for the output side CM current, 10-dB reduction can be observed when adopting the SW method. Therefore, the effectiveness of the proposed SW method is verified.

10.6 Summary

In this chapter, a split-winding (SW) method is proposed to reduce the common-mode (CM) currents in non-isolated DC-DC converters at both the input and output sides simultaneously. The SW method involves splitting and redistributing the inductor windings at the input, ground and output sides of the converter, in order to cancel the CM noise. Specifically, for the SW buck converter, SW boost converter, SW SEPIC converter operating in voltage step-up mode and SW Zeta converter operating in voltage step-down mode, the total cross-sectional areas of the windings remain the same as their original circuit. Considering imperfect coupling between the split windings, the balance condition is derived. The proposed SW method is verified by a 100-W buck converter, where 10- to 20-dB reduction for the CM noise at the input side is achieved, and 10-dB reduction for the CM noise at the output side is realized, demonstrating the effectiveness of the SW method.

References

1. EN 55022, Limits and methods of measurement of radio disturbance characteristics of information technology equipment, European: European Norm Standard (2006)
2. S. Wang, P. Kong, F.C. Lee, Common mode noise reduction for boost converters using general balance technique. IEEE Trans. Power Electron. **22**(4), 1410–1416 (2007)
3. L. Xie, X. Ruan, Z. Ye, Equivalent noise source: An effective method for analyzing common mode noise in isolated power converters. IEEE Trans. Ind. Electron. **63**(5), 2913–2924 (2016)
4. D. Cochrane, D. Chen, D. Boroyevic, Passive cancellation of common-mode noise in power electronic circuits. IEEE Trans. Power Electron. **18**(3), 756–763 (2003)
5. R.W. Erickson, D. Maksimović, *Fundamentals of Power Electronics* (Kluwer Academic, Norwell, Mass, 2001), pp. 107 and 806
6. K. Oberdieck, J. Gossmann, A. Bubert, R.W. De Doncker, Common- and differential-mode separators including the FM-broadcasting band. PCIM Europe (2018) pp. 1–8
7. J. Stahl, D. Kuebrich, A. Bucher, T. Duerbaum, Characterization of a modified LISN for effective separated measurements of common mode and differential mode EMI noise, in *2010 IEEE Energy Conversion Congress and Exposition (ECCE)*, Atlanta, GA, USA (2010), pp. 935–941

Chapter 11
Reducing the Common-Mode Currents at the Input and Output Sides in DC-AC Inverter Systems

11.1 A Generalized CM Noise Model of Inverter Systems

DC-AC inverters are widely applied in renewable energy generation, motor drive, and etc. Normally, the input and output of DC-AC inverters do not share the same ground, and the common-mode (CM) voltage of inverters causes the CM current. The CM current flows between the CM impedance at the input and output sides, leading to the conducted electromagnetic interference (EMI) and leakage current in inverter systems.

Figure 11.1 shows a single-phase photovoltanic (PV) inverter system with common-mode (CM) parasitic capacitances, where, C_{in1} and C_{in2} is the DC-link capacitors, Q_1 ~ Q_4 are the power switches, L_f is the output filter inductor, C_f is the output filter capacitor, Z_{line} is the line impedance, v_g is the grid voltage, and Z_g is the grounding impedance. Besides, C_{PV} is the parasitic capacitance between the PV panel and the protective earth (PE), C_{p1} ~ C_{p4} are the parasitic capacitance between the drain of Q_1 ~ Q_4 and the PE via the heatsink, respectively, C_{p5} is the parasitic capacitance between the input voltage negative terminal and the PE. As the inverter generates CM voltage, the CM currents i_{CM_in} and i_{CM_out} at the input and output sides are created through the CM impedance, leading to the CM EMI.

Figure 11.2 shows a typical three-phase motor drive system with the CM parasitic capacitances, where, C_{in1} and C_{in2} are the DC-link capacitors, Q_1 ~ Q_6 are the power switches of the inverter, C_p is the parasitic capacitance between the input source and the PE. Besides, C_{p1} ~ C_{p6} are the parasitic capacitance between the drain of Q_1 ~ Q_6 and the PE via the heatsink, C_{p7} is the parasitic capacitance between the input voltage negative terminal and the PE. Likewise, the inverter generates the CM voltage, and CM currents i_{CM_in} and i_{CM_out} at the input and output sides are created through the CM impedance, leading to the CM EMI.

© The Author(s), under exclusive license to Springer Nature Singapore Pte Ltd. 2024
X. Ruan et al., *Conducted Electromagnetic Interference in Power Converters: Modeling, Prediction and Reduction*, CPSS Power Electronics Series,
https://doi.org/10.1007/978-981-99-9295-9_11

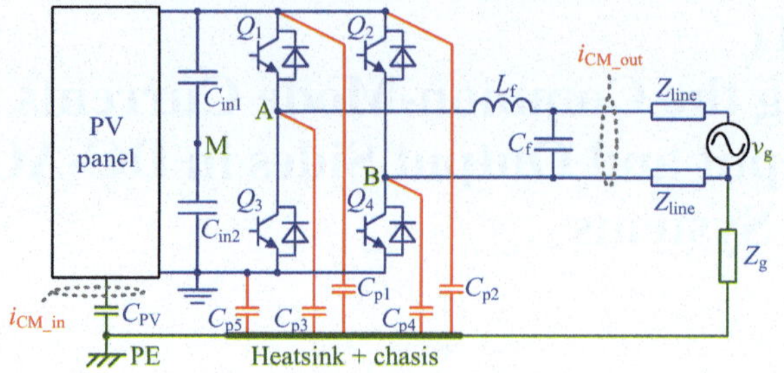

Fig. 11.1 PV inverter system with CM parasitic capacitances

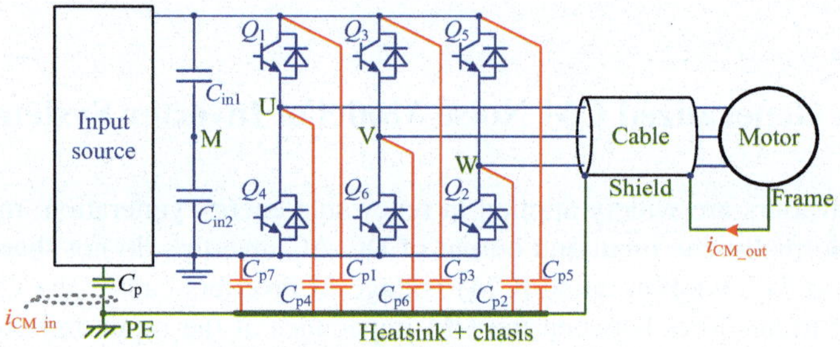

Fig. 11.2 Motor drive system with CM parasitic capacitances

By summarizing the CM noise models in PV inverter [1] and motor drive [2] systems, a generalized CM noise model is developed [2], as given in Fig. 11.3, where, Z_{CM_in} and Z_{CM_out} are the equivalent CM impedance of the input and output side, respectively. The voltage source is the CM voltage v_{CM} of the inverter that equals to the arithmetic average of the voltage across each phase-leg and the power ground. The capacitances C_{p_in} and C_{p_out} are related with the CM parasitic capacitances of the inverter. These parameters are given by

$$\begin{cases} v_{CM} = (v_A + v_B)/2 \\ C_{p\_in} = C_{p1} + C_{p2} + C_{p5} \quad \text{(for single-phase)} \\ C_{p\_out} = C_{p3} + C_{p4} \end{cases} \tag{11.1}$$

$$\begin{cases} v_{CM} = (v_U + v_V + v_W)/3 \\ C_{p\_in} = C_{p1} + C_{p3} + C_{p5} + C_{p7} \quad \text{(for three-phase)} \\ C_{p\_out} = C_{p2} + C_{p4} + C_{p6} \end{cases} \tag{11.2}$$

where, v_A, v_B, v_U, v_V and v_W are the voltages between each midpoint and the negative terminal of the input voltage, respectively.

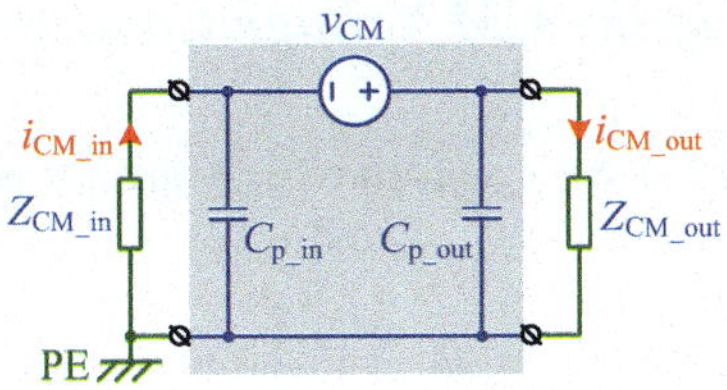

Fig. 11.3 Generalized CM noise model for single-phase and three-phase inverter systems

11.2 Derivation and Realization of the CM Voltage Cancellation Method

11.2.1 Derivation of the CM Voltage Cancellation Method

Referring to Fig. 11.3, the network consisted of v_{CM}, C_{p_in} and C_{p_out} with Δ-connection can be equivalently transformed into Y-connection [3] by applying the generalized Thevenin's theorem, as shown in Fig. 11.4a, where, C_{sum} and k_{inv} are expressed as

$$C_{sum} = C_{p\_in} + C_{p\_out} \tag{11.3}$$

$$k_{inv} = C_{p\_out}/C_{sum} \tag{11.4}$$

As seen in (11.3) and (11.4), k_{inv} is within (0, 1).

According to Fig. 11.4a, for counteracting the effect of two voltage sources, two cancellation voltage sources v_{com1} and v_{com2} can be introduced, as shown in Fig. 11.4b. When $i_{CM\_in} = i_{CM\_out} = 0$, v_{com1} and v_{com2} are expected to be

$$v_{com1} = k_{inv} v_{CM} \tag{11.5}$$

$$v_{com2} = (1 - k_{inv}) v_{CM} \tag{11.6}$$

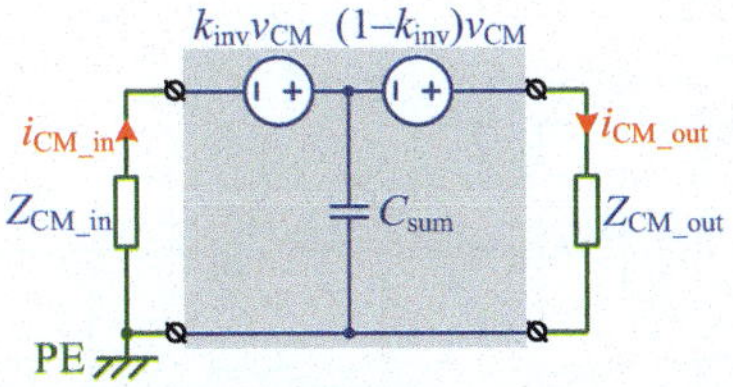

(a) The transformed equivalent circuit

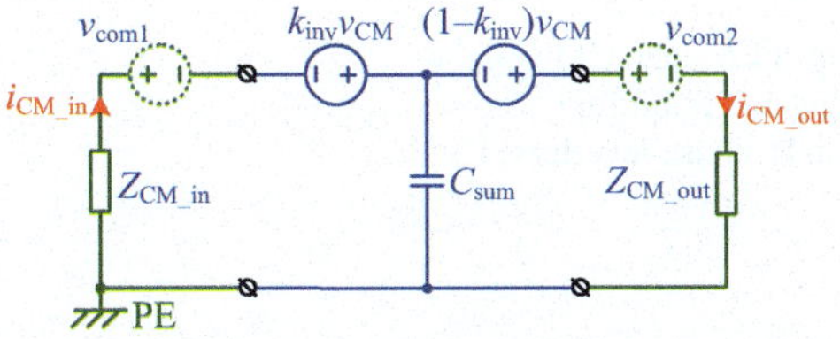

(b) The cancellation principle

Fig. 11.4 The equivalent CM noise circuit and the principle of cancellation

11.2.2 Realization of the CM Voltage Cancellation Method

According to (11.5) and (11.6), v_{com1} and v_{com2} are proportional to v_{CM}. So, it is necessary to sense v_{CM}, and inject v_{com1} and v_{com2} into the input and output sides of the inverter. Referring to Fig. 11.4b, Fig. 11.5 shows the principle of realizing the CM voltage cancellation (CMVC) method in inverter systems, where, the CM voltage sensing circuit senses v_{CM}, and the compensation voltage injection circuit generates v_{com1} and v_{com2} and injected them into the input and output sides, respectively.

Take the single-phase inverter system as an example, according to (11.1), the CM voltage v_{CM} is equal to $(v_A + v_B)/2$. In order to obtain $(v_A + v_B)/2$, the filter inductor L_f is implemented with a symmetrical coupled inductor with equal turns, and the two coupled windings are placed at the upper and lower sides, respectively. The filter capacitor C_f is divided into two capacitors C_{f1} and C_{f2}, and $C_{f1} = C_{f2} = 2C_f$, as shown in Fig. 11.6. Thus, we have $v_{Lf1} = v_{Lf2}$ and $v_{Cf1} = v_{Cf2}$. Owing to the symmetry, v_{MN} is expressed as

$$v_{MN} = \frac{V_{in}}{2} - \frac{(v_A + v_{Lf} - v_{Cf1}) + (v_B - v_{Lf} + v_{Cf2})}{2} = \frac{V_{in}}{2} - v_{CM} \quad (11.7)$$

where, V_{in} is the input DC voltage. As seen, the AC component in v_{MN} is $-v_{CM}$, so v_{MN} can be adopted to provide the CM voltage of the inverter.

Figure 11.7 highlights the circuit realization which adopts the CM transformer (CMT) for injecting the compensation voltage [4–6]. The sensed voltage v_{MN} is applied across the sensing windings ab of the CM transformer (CMT) T_{CM1} and

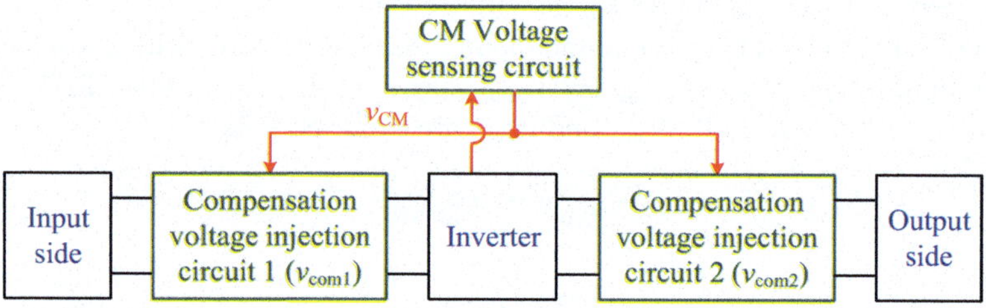

Fig. 11.5 The principle of realizing the CMVC method

Fig. 11.6 The CM voltage sensing circuit for single-phase inverters

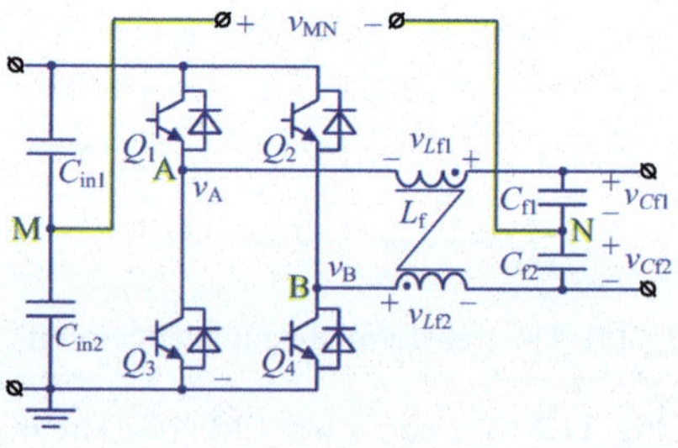

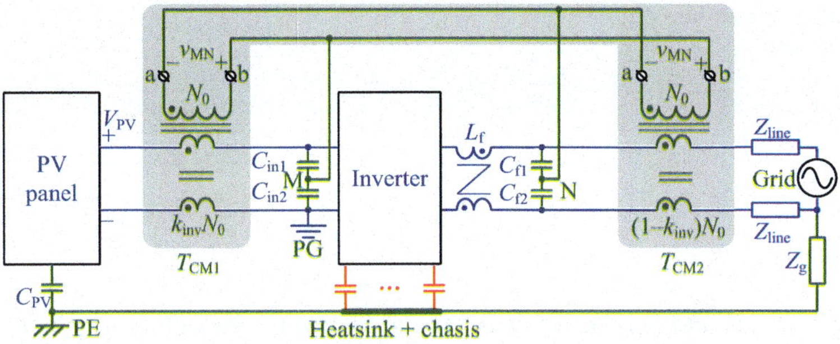

Fig. 11.7 The PV inverter system with CMT at both sides for CM current reduction

T_{CM2}, and the compensation voltage is induced across the injecting windings of T_{CM1} and T_{CM2}. Denoting the winding turns of ab in T_{CM1} and T_{CM2} as N_0, according to (11.5) and (11.6), the turns of the injecting windings in T_{CM1} and T_{CM2} should be $k_{inv}N_0$ and $(1–k_{inv})N_0$, respectively. For three-phase inverter systems, the realization of the CMVC method is similar, and it is not discussed here.

11.3 Practical Considerations for the CMVC Method in Inverter Systems

In this Section, the practical considerations for the CMVC method are discussed.

11.3.1 Restriction of Input and Output Currents

Since the injecting windings are inserted into the DC and AC sides, the proposed method is preferred in low-current applications.

11.3.2 Balancing Capacitor

In (11.4), k_{inv} is determined by the capacitance ratio. However, the parasitic capacitances take arbitrary value within a reasonable range. For making the turns $k_{inv}N_{ab}$ and $(1 - k_{inv})N_{ab}$ of two CMTs be an integer, a balancing capacitor could be added and connected in parallel with one of the parasitic capacitance.

When power switches adopts discrete MOSFETs, the dominant CM parasitic capacitances are the ones between the drain of each MOSFET and the heatsink. As long as the same insulation material with equal thickness is inserted between each MOSFET and the heatsink, these parasitic capacitances are equal. According

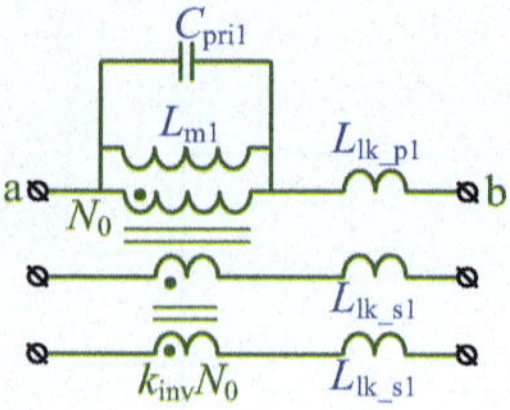

Fig. 11.8 The parasitics of the CMT T_{CM1}

to (11.1), (11.3) and (11.4), k_{inv} is 0.5 regardless of the capacitance. When the CM voltage sensing winding ab takes even turns, the turns of injecting windings will be an integer, and balancing capacitor is not required.

11.3.3 The Influence of Parasitics in the CMT

The CMT is a critical component for sensing and injecting the CM voltage, and its parasitics affect the CM current reduction at high frequencies. Figure 11.8 shows T_{CM1} with parasitics, where L_{m1} is the magnetizing inductance, C_{pri1} is the parasitic capacitance of the sensing winding, $L_{\text{lk\_p1}}$ and $L_{\text{lk\_s1}}$ are the leakage inductance of the sensing winding and the injecting windings, respectively. The leakage inductance between the injecting windings is neglected as they are tightly coupled.

At high frequencies, the influence of C_{pri1} and $L_{\text{lk\_p1}}$ is dominant, and the induced voltages of the injecting windings are no longer exactly proportional to v_{CM}. In order to achieve a wide frequency range for reducing the CM currents, C_{pri1} and $L_{\text{lk\_p1}}$ should be kept small, and sandwiched winding structure is preferred to realize the CMT.

11.3.4 Comparison with Other Methods

The passive filter, floating filter [7], active cancellation method [4] and the CMVC method in this chapter are compared in terms of their used components, bandwidth and noise attenuation, as shown in Table 11.1. The implementations for these existing methods are shown in Fig. 11.9. In Fig. 11.9a, for the passive filter approach, L_{CM1} and L_{CM2} are the CM inductors at the input and output sides, C_{y1} and C_{y2} are the CM capacitors at the input and output sides respectively. For the floating filter in Fig. 11.9b, the CM capacitors C_{y1} and C_{y2} are connected to the midpoint of the DC-link capacitors, which aims to decouple the input and output CM current [7]. Figure 11.9c shows the implementation of active cancellation method, where, the CM voltage of the inverter is sensed and sent into a power amplifier, and the compensation voltage, which equals to v_{CM}, is injected into the output through a CMT to counteract the effect of the CM voltage of the inverter.

Table 11.1 Comparison between passive filter, floating filter, active cancellation and the CMVC methods

		Passive filter	Floating filter	Active cancellation	CMVC method
Components	Active	0	0	2 power transistors	0
	Passive	2 CM inductors, 4 CM capacitors	2 CM inductors, 4 CM capacitors	1 CMT	2 CMTs
Bandwidth	f_L	f_L is designed to be far lower than the switching frequency f_s			
	f_H	f_H is limited by the EPC of CM inductors and the ESL of CM capacitors		f_H is limited by the leakage inductance and parasitic capacitance of the CMT	
Noise attenuation	ac side	If Z_S and Z_L are far lower than the impedance of CM inductors in Fig. 11.9a, b, strong attenuation of CM current can be achieved. Otherwise, the attenuation is not significant		Yes, strong	Yes, strong (40 dB for the experiment in this chapter)
	dc side			Yes, Medium (20 dB in [4])	

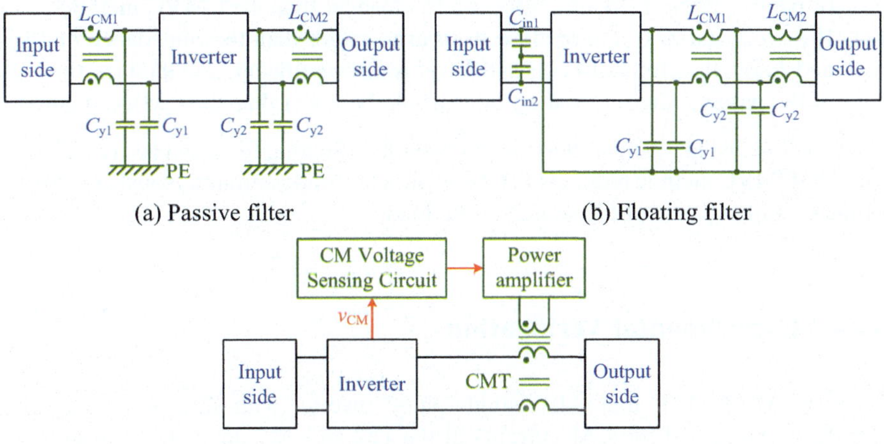

(a) Passive filter (b) Floating filter (c) Active cancellation

Fig. 11.9 The implementations for the existing methods

For these four methods, the effective frequency range is denoted as $[f_L, f_H]$. The lower cut-off frequency f_L for these methods is required to be much lower than the switching frequency, which can be realized by choosing proper parameters. The higher cut-off frequency f_H of passive filter and floating filter is limited by the equivalent parallel capacitance (EPC) of the CM inductors and the equivalent series inductance (ESL) of the CM capacitors. For the active and the CMVC method, f_H is limited by the leakage inductance and parasitic capacitance of the CMT.

Based on the experimental results in [4] and this chapter, the AC-side CM current is significantly reduced by both methods. For the DC-side CM current, the active cancellation gives a reduction of 20 dB, and the CMVC method reaches a reduction of 40 dB. Because two CMTs are used in the CMVC method, the volume of passive components is larger than that in the active method. However, as the power transistors in the active cancellation method need cooling parts, the volumetric density for the inverter system with these two methods depends on the applications.

For a fair comparison of the noise attenuation among the passive filter, floating filter and the CMVC method, the component volume of the CM inductors or CM transformers, which take up most of the volume in these methods, are kept the same. Therefore, the CM inductances of L_{CM1} and L_{CM2} in Fig. 11.9a, b are assumed to be equal to the magnetizing inductance of the CMT in Fig. 11.9c at the input and output side in the CMVC method, respectively. Because the noise attenuation of passive filter and floating filter is affected by the CM impedance of the source and the load, the noise attenuation of these methods varies with the situations:

(1) If Z_{CM_in} and Z_{CM_out} are far lower than the impedance of L_{CM1} and L_{CM2}, L_{CM1} and L_{CM2} will sustain most of the CM voltage of the inverter, and the attenuation of passive filter or floating filter can be stronger than the CMVC method.
(2) If Z_{CM_in} and Z_{CM_out} are close to or even larger than the impedance of L_{CM1} and L_{CM2}, the effect of L_{CM1} and L_{CM2} is not significant, and the insertion loss of the passive filter or floating filter would be lower than the CMVC method.

As a conclusion, compared to the passive filter, floating filter and active cancellation, the CMVC method reduces the CM noise at both sides simultaneously regardless of the CM impedance of the source or the load.

11.4 Experimental Verification

To verify the CMVC method, a single-phase inverter with silicon-carbide (SiC) devides is built, and the CM currents at the DC and AC sides are measured and compared.

11.4.1 Specifications of the Prototype

Figure 11.10 shows the single-phase inverter prototype and the output filter. The power switches $Q_1 \sim Q_4$ adopt SiC MOSFETs (CREE: C2M0040120D), and the output filter capacitor uses C_{f1} and C_{f2} in series. Figure 11.11 shows the measurement setup for the CM current. As seen, two capacitors C_1 (C_2) are connected between

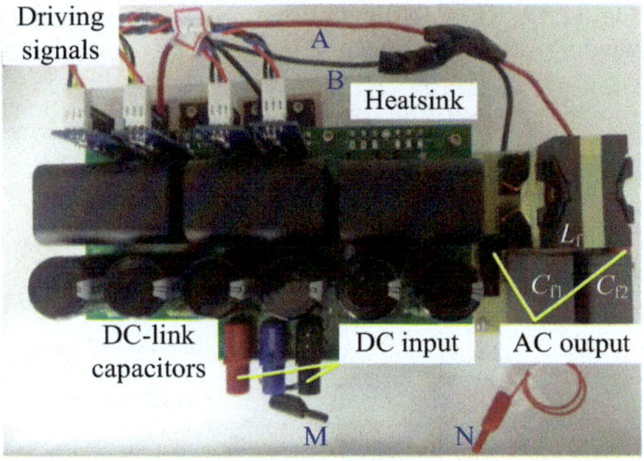

Fig. 11.10 Photo of the SiC inverter prototype

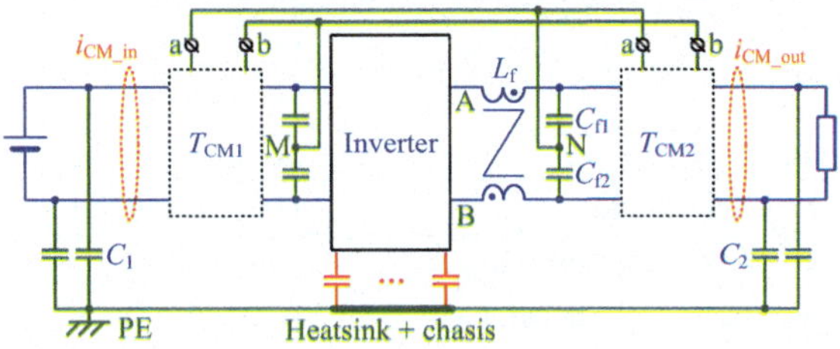

Fig. 11.11 Measurement setup for the CM current

the input power cords (loads) and the PE, serving as the CM impedances at the DC and AC sides.

Table 11.2 lists the parameters and the CM parasitic capacitances of the inverter. The parasitic capacitance C_p between the drain of each MOSFET and the heatsink is measured by an impedance analyzer (WAYNE KERR 6500B) with MOSFETs disconnected from the inverter, and the test frequency is 100 kHz.

Figure 11.12 shows the photo and winding structure of T_{CM1} and T_{CM2}. Sandwiched winding structure is adopted to reduce the leakage inductance and equivalent parasitic capacitance in parallel with the sensing winding of T_{CM1} and T_{CM2}. Table 11.3 lists the parameters of the magnetic component L_f, T_{CM1} and T_{CM2}. Since

Table 11.2 The main circuit parameters and the CM parasitic capacitances

DC input voltage	400 V	DC-link capacitance	220 μF
Load resistance	50 Ω	Filter inductance L_f	520 μH
Nominal power	1 kW	Filter capacitance C_b	6.8 μF
Output frequency	50 Hz	Parasitic capacitance C_p	38 pF
Switching frequency	80 kHz	Input side capacitance C_1	0.1 μF
Modulation	Unipolar SPWM	Output side capacitance C_2	1 nF

Fig. 11.12 The CMT and its winding structure

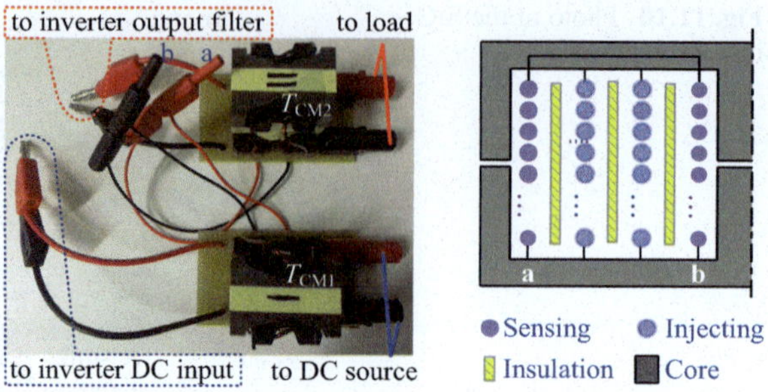

(a) Photo of the CMT (b) Winding structure

Table 11.3 The parameters of the magnetic components

	L_f	T_{CM1}	T_{CM2}
Core	PQ 50/50	PQ 40/40	
Turns	22: 22	30: 15: 15	
Magnetizing ind	130 μH, 130 μH	3.13 mH	3.32 mH
Primary leakage ind. L_{lk_p}		1.30 μH	1.36 μH
Secondary leakage ind. L_{lk_s}		0.48 μH	0.46 μH
Parasitic capacitance C_{pri}		27.4 pF	28.1 pF

discrete MOSFFETs are used, the equivalent parasitic capacitance C_{p_in} and C_{p_out} are both equal to $2C_p$ (38 pF), and the turns ratio of T_{CM1} and T_{CM2} are both 2:1:1 ($k_{inv} = 0.5$) according to (11.4).

To extract the leakage inductance and the parasitic capacitance of the CMT, the impedance curve of each winding in the CMT are measured. Based on the impedance curve, the calculated parameters are also listed in Table 11.3. The resonant frequency between L_{lk_p} and C_{pri} of T_{CM1} and T_{CM2} is 26.67 MHz and 25.75 MHz, respectively.

11.4.2 The Verification of the CMVC Method

Figure 11.13 shows the measured voltages for the circuit in Fig. 11.11 at full-load, where, v_{CM} is the measured CM voltage across points M and N, and the injected voltages v_{TCM1} and v_{TCM2} are measured across the injection windings of T_{CM1} and T_{CM2}, respectively. As seen, their waveforms are generally similar, and have amplitude ratio of 2:1:1 as expected.

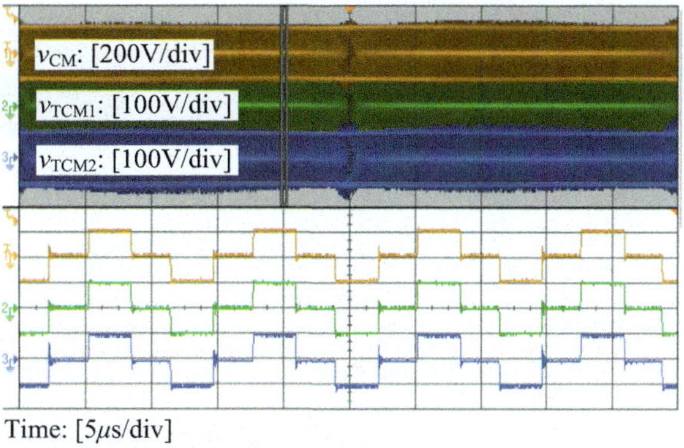

Fig. 11.13 The CM voltage of the inverter and the injected voltage from the CMT

To verify the effectiveness of the CMVC method, i_{CM_in} and i_{CM_out} in Fig. 11.11 are measured with high-bandwidth current probes (Agilent 1147A) under three cases, described as follows.

Case I: The original case without T_{CM1} and T_{CM2};

Case II: T_{CM1} and T_{CM2} are inserted with leaving their sensing windings ab open (T_{CM1} and T_{CM2} act as CM inductors);

Case III: T_{CM1} and T_{CM2} are inserted with the connection according to Fig. 11.11 (T_{CM1} and T_{CM2} act as the CMT in this case).

Figure 11.14 shows the time-domain waveforms of the currents i_{CM_in} and i_{CM_out} under the aforementioned three cases. It is observed that the original case without the CM inductors or the CMVC method has the highest amplitude.

As show in Fig. 11.14a, for Case I, i_{CM_in} and i_{CM_out} are similar. According to the CM noise model in Fig. 11.3, Z_{CM_in} and Z_{CM_out}, which equal to the impedance of $2C_1$ and $2C_2$ in the experiment, are far smaller than the impedance of C_{p_in} and C_{p_out}. Therefore, C_{p_in} and C_{p_out} can be treated as open-circuit, and the simplified circuit indicates $i_{CM\_in} = i_{CM\_out}$.

As shown in Fig. 11.14b, for Case II, where the CM inductors are inserted, the amplitude of i_{CM_in} and i_{CM_out} is significantly reduced. However, the harmonic amplitude in i_{CM_in} and i_{CM_out} at the switching frequency can still be observed, as highlighted in Fig. 11.14b. This is because the impedance of the CM inductors is small at the switching frequency, and the attenuation of i_{CM_in} and i_{CM_out} is not significant.

Figure 11.14c shows the measured results for Case III. As seen, i_{CM_in} and i_{CM_out} are further reduced with the CMVC method. Note that there are little spikes in i_{CM_in} and i_{CM_out}, because the parasitic capacitance between Q_1 ~ Q_4 and the PE is not exactly the same, and the CM noise cancellation is not perfect as expected.

Figure 11.15 compares the CM current spectra of i_{CM_in} and i_{CM_out} under the three cases. The CM current data is collected by a digital oscilloscope (Agilent MSO-X 3054A), and then sent to MATLAB for calculating the CM current spectra using fast Fourier transform (FFT) with rectangular window. Normally, the time length for the

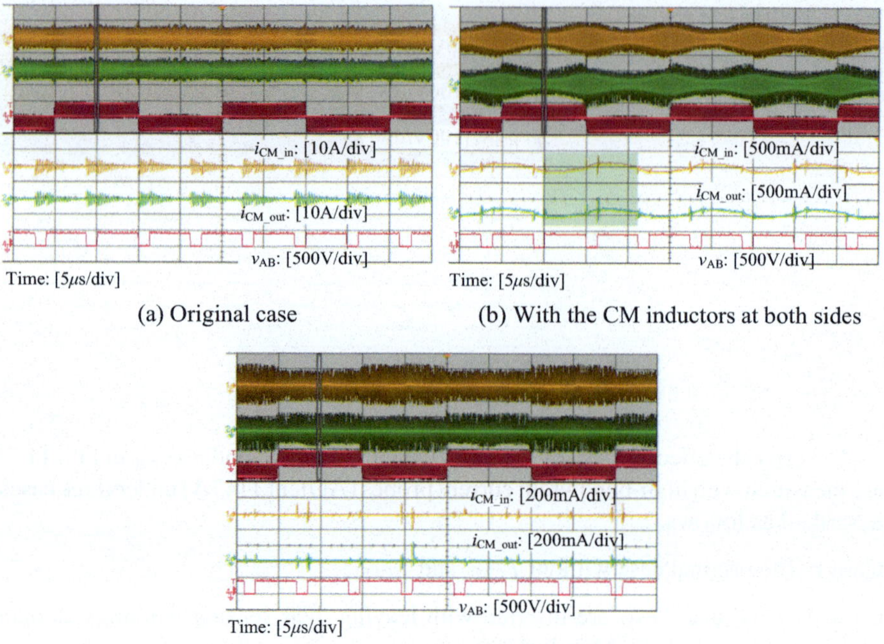

(a) Original case (b) With the CM inductors at both sides

(c) With the CMT at both sides

Fig. 11.14 Comparison of CM currents at the DC and AC sides in the time-domain

FFT should be 20 ms for covering a complete line cycle, but the calculated spectrum only ranges from 0 to 1.56 MHz due to limited sampling points of the oscilloscope. To observe the high frequency components, a 50-µs time length is adopted for the FFT to extend the higher frequency range, and the CM current data are recorded when the output voltage is crossing zero.

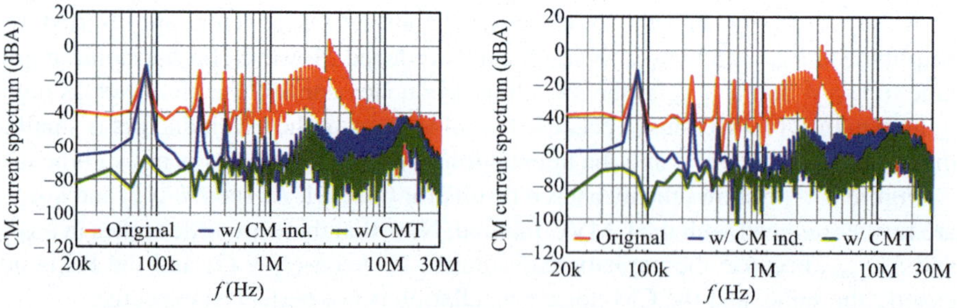

(a) The CM current spectrum at the DC side (b) The CM current spectrum at the AC side

Fig. 11.15 Comparison of CM current spectra at DC and AC sides with a 50-µs time length

As seen in Fig. 11.15, the harmonic spectra of i_{CM_in} and i_{CM_out} for the original case (Case I) are the highest, and there are clear spikes at the switching harmonics. After inserting the CM inductors (Case II), the harmonic amplitude of i_{CM_in} and i_{CM_out} see a reduction at high frequencies. However, the harmonic amplitude around the switching frequency is still high, because the impedance of CM inductors is not sufficiently larger than that of C_1 and C_2. Finally, by applying the CMVC method (Case III), the spectra of i_{CM_in} and i_{CM_out} are further reduced. Compared to the original results, nearly 40-dB reduction from 20 kHz to 10 MHz is achieved by adopting the CMVC method. The higher cut-off frequency for effective cancellation is 10 MHz, which is nearly 1/3 of the resonant frequency between L_{lk_p} and C_{pri} of T_{CM1} and T_{CM2}. Thus, the proposed CMVC method is verified.

11.5 Summary

In this chapter, the CMVC method for inverter systems is investigated, which reduces the CM currents at both the input and output side. The CMVC method for inverter system consists of the CM voltage sensing and injection circuits. For a single-phase inverter, the CMVC method requires a differentially coupled filter inductor and two capacitors for sensing the CM voltage, and two CM transformers for injecting the compensation voltage. A single-phase inverter prototype with silicon-carbide (SiC) devices is built in the lab, and the measured upper frequency for the CM current reduction is 10 MHz. By implementing the CMVC method, 40-dB reduction for the CM current spectra at both the input and output sides is achieved, verifying the effectiveness of the CMVC method in inverter systems.

References

1. H. Xiao, S. Xie, Leakage current analytical model and application in single-phase transformerless photovoltaic grid-connected inverter. IEEE Trans. Electromagn. Compat. **52**(4), 902–913 (2010)
2. S. Wang, Y.Y. Maillet, F. Wang, D. Broyevich, R. Burgos, Investigation of hybrid EMI filters for common-mode EMI suppression in a motor drive system. IEEE Trans. Power Electron. **25**(4), 1034–1045 (2010)
3. M.F. Moad, Two-port networks with independent sources. Proc. IEEE **54**(7), 1008–1009 (1966)
4. S. Ogasawara, H. Ayano, H. Akagi, An active circuit for cancellation of common-mode voltage generated by a PWM inverter. IEEE Trans. Power Electron. **13**(5), 835–841 (1998)
5. M.M. Swamy, K. Yamada, T. Kume, Common mode current attenuation techniques for use with PWM drives. IEEE Trans. Power Electron. **16**(2), 248–255 (2001)
6. Y. Murai, T. Kubota, Y. Kawase, Leakage current reduction for a high-frequency carrier inverter feeding an induction motor. IEEE Trans. Ind. Appl. **28**(4), 858–863 (1992)
7. Y. Liu, Z. Mei, S. Jiang, W. Liang, Conducted common-mode electromagnetic interference suppression in the AC and DC sides of a grid-connected inverter. IET Power Electron. **13**(13), 2926–2934 (2020)

SPRINGER NATURE

GPSR Compliance

The European Union's (EU) General Product Safety Regulation (GPSR) is a set of rules that requires consumer products to be safe and our obligations to ensure this.

If you have any concerns about our products, you can contact us on ProductSafety@springernature.com

In case Publisher is established outside the EU, the EU authorized representative is:

Springer Nature Customer Service Center GmbH
Europaplatz 3
69115 Heidelberg, Germany

The manufacturer's authorised representative in the EU is Springer Nature Customer Service Centre GmbH, Europaplatz 3, 69115 Heidelberg, Germany. If you have any concerns regarding our products, please contact ProductSafety@springernature.com

Printed and bound by CPI Group (UK) Ltd, Croydon, CR0 4YY

01/10/2026

02225868-0002